PRAISE FOR
CREATING LOYALTY IN YOUTH

'A revealing look at the $400 billion youth travel market which has doubled in size since 2009 and continues to grow. Author Stephen Lowy, a UK-based travel executive, includes interviews with leaders from the airline, hostel, hotel and tourism and tour operator segments. This is essential reading for those in and out of this field, with future trends and opportunities explored in detail.'
William L Gertz, Chairman, American Institute for Foreign Study (AIFS)

'Passionate and insightful. Stephen Lowy and his colleagues offer perceptive ideas on the historical evolution, current status and future trends in youth travel. From personal and professional knowledge, they highlight youth travel's contribution to economic development, and the impact on participants' lifelong learning. Their perspectives on marketing and programme design make this a must-read for individuals engaged in innovative, experiential global learning enterprises.'
Yuhang Rong, Strategic Adviser for North America, Kaplan International Pathways

'Demonstrates how the youth travel market is both incredibly valuable and varied, reflecting the diverse motivations of young travellers and the broad range of often life-changing experiences from across the globe. The book eloquently explains the evolution of youth travel, offering essential context for understanding its present state, making it necessary reading for anyone operating in this space, adamant about being a part of it in the future.'
Niclas Lundquist, Head of Commercial (UK), KILROY, Denmark

'A must-read for anyone interested in understanding how youth travel can drive long-term economic prosperity, cultural understanding and brand loyalty. Full of unique insights and practical tips on how to attract, build and maintain relationships with young travellers, this gem of a book is an invaluable resource for tourism policy makers, business leaders and marketeers.'
Angela Maher, Chief Executive, Savoy Educational Trust

'Stephen Lowy is one of very few who truly understand youth travel, a diverse and fast-changing industry. He generously shares decades of experience of being an adventurous traveller and hands-on professional, touching on most functions, from operations to sales, marketing and management. Few people have the insight he has and truly understand the business. To be successful in this industry, with ever-increasing acquisition costs, it is key to understand what drives loyalty and repeat business.'
Anders Ahlund, President of International Sales and Marketing, EF Education First

'In a global environment that is becoming increasingly sceptical of multilateralism and where governments are looking inwards rather than outwards, the need for citizens to citizen diplomacy has never been more important.'
Dr. Paul Beresford-Hill CBE, Ambassador and Permanent Observer, United Nations

Creating Loyalty in Youth Travel

*How travel brands and destinations
can build lifelong relationships*

Stephen Lowy

KoganPage

First published in Great Britain and the United States in 2025 by Kogan Page Limited

Kogan Page
Kogan Page Ltd, 2nd Floor, 45 Gee Street, London EC1V 3RS, United Kingdom
Kogan Page Inc, 8 W 38th Street, Suite 902, New York, NY 10018, USA
www.koganpage.com

EU Representative (GPSR)
Authorised Rep Compliance Ltd, Ground Floor, 71 Baggot Street Lower, Dublin D02 P593, Ireland
www.arccompliance.com

Kogan Page books are printed on paper from sustainable forests.

ISBNs
Hardback   978 1 3986 2064 3
Paperback   978 1 3986 2063 6
Ebook   978 1 3986 2062 9

British Library Cataloguing-in-Publication Data
A CIP record for this book is available from the British Library.

Library of Congress Control Number
2025009712

Typeset by Integra Software Services, Pondicherry
Print production managed by Jellyfish
Printed and bound by CPI Group (UK) Ltd, Croydon CR0 4YY

CONTENTS

Foreword x
Acknowledgements xiii

Introduction 1

1 What is youth travel? 6
The evolution of youth travel 6
Government policies and youth travel 7
Marketing youth travel: The Millennial and Gen Z divide 8
Educational benefits of youth travel 8
The post-Covid phenomenon: 'Revenge travel' 9
A broad definition of youth travel 9
Notes 10

2 The value of the youth travel market 11
Introducing Bonard and its role in youth travel research 11
The evolution of youth and student travel over 15 years 12
The impact of technology on youth travel trends and best practices in
    data utilization 12
Post-Covid recovery in youth and student travel 14
Youth travel's economic contributions 16
The long-term value of youth travel 17
Conclusion 21
Key takeaways 21
Notes 23

3 Attracting the youth travel market 24
The growing power of youth travel 24
Resilience in action: Lessons from Covid and beyond 28
Marketing to Gen Z: Authenticity, humour and digital first 30
Creating a strategy that lasts: Aligning marketing, product
    and policy 33

Conclusion 37
Key takeaways 38
Notes 38

4   Building community 39
From backpacker budget to luxury experiences 39
The transformation of hostels 40
Changing needs of today's youth travellers 41
Youth-centric destinations: The standouts 42
The importance of infrastructure for youth travel 45
Opportunities for smaller destinations 45
Future trends in youth travel 46
Planning for youth travel success 49
Conclusion 50
Key takeaways 51

5   Investing in youth travellers as key drivers of economic
    growth 53
The journey from Working Holiday visa to economic driver 54
Media and popular culture: Shaping perceptions 55
Youth travel: A key component of Australia's economy 55
The long-term value of youth travellers 56
Lessons from a long-haul destination 57
Creating lifelong ambassadors 62
Conclusion 65
Key takeaways 65

6   Authenticity, safety and growth 67
Trust and safety as competitive advantages 68
Authentic experiences drive success 69
Collaborations and ecosystems 73
The digital evolution of youth travel 76
Innovation is key for youth 82
Conclusion 86
Key takeaways 86
Notes 87

7   Challenges facing the youth travel sector and what they mean for the tourism industry 88

BETA and its role in youth travel 88
Founding BETA: A unified voice for the sector 89
Quantifying the sector: The power of data 91
The UK government's role: Understanding and missteps 92
Global comparisons: Learning from the best 94
Entrepreneurship in the sector 96
Conclusion 97
Key takeaways 98

8   Maintaining brand loyalty 100

Attracting younger demographics through targeted brands 101
Leveraging loyalty programmes for lifetime value 102
Engaging customers beyond stays 104
Social media and the rise of TikTok 106
Strategic investment in digital transformation 109
Aligning global presence with local relevance 110
The importance of customer lifetime value 111
Conclusion 112
Key takeaways 112

9   Destination marketing and management best practices 114

Understanding Toposophy: A place-based approach to tourism
    strategy 114
Youth travel: A tool for balancing over-tourism 115
Tourism ecosystems: The interconnected framework of a
    destination 117
Greece's potential in youth travel 118
The impact of politics and policy on tourism 118
Building long-term affinities through youth travel 119
Challenges and opportunities in youth travel 121
Athens: An emerging hub for digital nomads 122
Data-driven tourism strategies: A game changer 124
Conclusion 125
Key takeaways 125
Note 126

10 Language travel 127

The dual role of youth travel: Bridging education and tourism 128
From short courses to long-term journeys 129
Planting the seed: The early impact of youth travel 132
Connecting the dots: The disjointed ecosystem of youth travel 133
The value of in-person experiences 137
The multiplier effect of youth travel 143
Conclusion 144
Key takeaways 145
Notes 145

11 The lifelong value of youth mobility 146

The skills and lessons of travel 147
The role of policy and programmes 150
Economic and cultural impact 153
Challenges and opportunities 155
Conclusion 163
Key takeaways 164

12 Reinventing hostels for a new generation 166

Hostelworld's role in the evolving youth travel market 167
The changing face of hostel guests 168
APAC's growing hostel landscape 169
The rise of digital-first travel 169
Understanding Asian traveller preferences 170
Leveraging marketing to target Asian travellers 172
Balancing technology and authentic experiences 177
Mental health and travel: Disconnecting to reconnect 178
Conclusion 181
Key takeaways 181

13 Redefining youth living 183

Introduction to youth-focused living 183
Generational trends shaping youth travel 185
The importance of long-term planning 188
The role of feedback in innovation 189

From students to professionals 193
Conclusion 196
Key takeaways 196

14 The shifting landscape of youth marketing 198

Redefining youth: The challenges of broad age brackets 199
Lessons from the field: Key success factors in campaigns 200
Youth as influencers: Beyond direct ROI 201
Striking the balance: Short-term wins vs long-term gains 201
Adapting to change: Learning from sectors outside travel 203
The technological divide: Navigating platform preferences 206
Evolving traveller segments: Tech-driven vs tech-free explorers 209
Conclusion 211
Key takeaways 212

15 Engaging the next generation 213

The role of tourism in a global football brand 213
Engaging the youth demographic 215
Creating unforgettable visitor experiences 217
Social media and the youth market 219
Collaboration with Liverpool city region 220
A blueprint for youth-centric tourism 223
Conclusion 223
Key takeaways 224

16 The future of youth travel 225

Challenges and opportunities for the future 225
The road ahead: Predictions and strategies 227

Index 229

FOREWORD

The opportunity for young people to travel the world seems so obvious to us now, a natural extension of exploration and connection. However, in the 1970s, this was far from the reality for most. The idea of leaving your home country to visit another was often perceived as an exotic and daunting step into the unknown. While the 'Hippy Trails' – paths from first-world countries to destinations like India – were beginning to attract a niche group of adventurers, the concept of international travel for the masses, especially for young people, remained largely uncharted territory.

World War II (WW2) was a stark reminder of the devastating consequences of division and conflict. In its aftermath, a profound thought emerged: could the mobility of young people, exploring and engaging with other countries, help prevent such tragedies from recurring?

The vision was simple but powerful. By encouraging youth to travel internationally, to experience different cultures first-hand, and to form meaningful connections with people from diverse backgrounds, the hope was to foster greater understanding and empathy. These young travellers could return home with enriched perspectives, armed with lessons and insights that might bridge divides and contribute to a more harmonious global community.

When WW2 ended in 1945 the International Air Transport Association (IATA) was formed and, as a part of the vision to avoid future wars, students were left out of the restrictive regulations (together with military, missionary and marine sectors, all for various practical policy reasons), with the hope that this would encourage young people to explore the world and bring down cultural barriers. In Europe, the European Civil Aviation Conference (ECAC) controlled charters which had an ECAC rule 6 (later renumbered ECAC 3) which allowed specialist operators to operate charters only for eligible full-time students who could buy one-way tickets (all other charters required passengers to buy return tickets). This again encouraged the mobility of students to learn about and explore the world with flexible plans. Similar regulations existed in other countries and continents.

The Student Air Travel Association (SATA) was set up for specialist student agencies to collaborate on ECAC 3 flights, meaning the departure and arrival countries could sell into charters, thus spreading the risk and benefit. The International Student Identity Card (ISIC) was the means by which full-time eligible students could travel – linked to UNESCO.

The charter flights were booked through the registered members of SATA, with a SATA administration office in Copenhagen, and bookings were made by telex and ticketed on manually produced two or four coupon tickets. Flight operators collaborated so that, for example, a number of charters would fly to one city from a number of cities and then a limited number would get onto another flight to go long haul, or link to other short haul charters. So, for example, flights from some big European cities would fly to Copenhagen and then some from each flight would get on another aircraft to Bombay while most would stay in Copenhagen or would go onwards to the originating city of other feeder flights. Complicated by today's standards!

The big change came in the early to mid 1970s with the 1973 Yom Kippur War, which sent the oil price from 90 cents to $15 overnight. This sent over 90 per cent of the European and world travel industry into liquidation very quickly (private companies, but not airlines as these were generally government owned or backed, requiring huge government support to continue). New investors moved into the space and many of the names that dominated are still in operation today (such as TUI, etc).

The transformative shift brought about by the new startups in the student and youth retail travel sector was their decision to employ a new generation of travellers. These were individuals who had first-hand experience of exploration, who had visited the destinations and lived through the journeys they were now promoting.

Through the 1970s these student specialist agencies continued to charter, but also began to collaborate, firstly with non-International Air Transport Association (IATA) and later with IATA airlines, to accept their student 'charter' ticket coupons on their scheduled flights. This required real sophistication in terms of managing paper flow and accounting as well as the business model (a four-coupon ticket valid for up to 12 months and each coupon manually processed) and financial management and security.

In the 1980s ticket eligibility was extended in some cases to include non-students under 26 years of age. The concept of young travellers exploring the world for their short- and long-term gain was born. This was enhanced with bilateral working holiday and study visas between many countries. International organizations, the International Student Travel Confederation (ISTC) and the Federation of International Youth Travel Organizations (FIYTO), became the focal points for those engaged in the trade.

Of course, there was a bigger effect in the supply chain, with accommodation providers and the broad leisure industry targeting the young travellers becoming a fast-growing business full of invention and opportunity. That included specialists but also mainstream suppliers of accommodation, tours, catering, etc.

Young people travelling internationally was also seen by many governments as an opportunity – both having young people arriving internationally to fill vacancies, and also their own young people seeing the world and returning with more of a global outlook, to the long-term benefit of their countries.

Another outcome from the 1980s was the opportunity for many countries to expand the educational opportunities for their own people, and large numbers of young people travelled to study. The expansion for universities and language schools, together with businesses looking for talent, was an enormous growth area which has continued since that time. The transformative effect on many second and third world countries has been profound as their young people internationalized their experiences and returned to change their countries forever.

Many of the iconic travel brands of the 1970s and 1980s have faded into history, yet the core principles and innovations they introduced continue to shape modern travel, education and global societies. These pioneering ideas – designed to inspire exploration, foster cultural exchange and create transformational experiences – remain as relevant today as they were decades ago. Today's adventurers continue to retrace and reinvent the journeys of their parents and grandparents, proving that the spirit of discovery transcends time, adapting to new contexts while remaining rooted in timeless principles of connection and curiosity.

For those reading this book, whether you're a policymaker, a business leader or simply someone passionate about travel, I hope this foreword inspires you to think about the broader impact of youth mobility. It's not just a sector of the travel industry – it's a force for global understanding, innovation and progress. Let us continue to champion the spirit of exploration, ensuring that the opportunities for young people to see the world remain accessible for generations to come.

Dick Porter
Former CEO, STA Travel

ACKNOWLEDGEMENTS

Writing this book has been an overwhelming yet deeply rewarding journey, one that I never imagined I would embark upon. For many years, I've been fortunate to work in the youth travel sector – a space filled with passionate and creative individuals united by the common goal of providing young people with life-changing experiences. This project gave me the opportunity to reflect on that shared mission, and I am incredibly grateful for the chance to shine a light on the people and organizations that make it possible.

It would be difficult to name everyone who has played a role in shaping my career, education and life. To anyone I may have missed mentioning by name, please know that your support and influence have not gone unnoticed. I am truly grateful for the many individuals who have supported, guided and inspired me throughout my journey. Your contributions, whether big or small, have left an indelible mark on my path, and for that I thank you.

First and foremost, my heartfelt thanks go to Fergus Boyd, whose words of encouragement planted the seed for this book. Fergus, your IT-related leadership in transforming some of the world's most renowned travel and hospitality brands has always inspired me, and your belief in my ability to write a book gave me the courage to start.

This book would not have been possible without the steadfast support of Susan Furber and Jeylan Ramis at Kogan Page. Balancing the demands of writing with my full-time role as a CEO, which inevitably involves significant travel (no surprises there!), was no small feat. Your patience, guidance and encouragement kept me going during moments of doubt and helped me bring this project to completion.

To the leaders in travel and hospitality featured in this book: thank you. Your time, passion and commitment to creating or supporting life-changing opportunities for young people are remarkable. What you do daily ensures that countless young travellers have transformative experiences. To everyone running businesses – large and small – that support youth travel, please keep doing what you're doing. Your work changes lives.

To my wife, Linda, you have been my rock throughout this process. Thank you for enduring the early mornings and weekends when I was glued to my keyboard. Your unwavering support – both for this project and throughout my career – has been my anchor. Our own story began because of youth travel, and it's a beautiful reminder of how deeply this sector is intertwined with our lives.

To my father, Peter, thank you for giving me the gift of travel at a young age and for encouraging me to pursue a career in this incredible industry. Your pioneering work in the 1970s, creating spaces for young travellers in London, laid the foundation for the thriving sector we see today. You've inspired not only me but also generations of travellers.

To my late mother, Elizabeth, or Lizzie as you were lovingly known, your support in everything I did, from sports to life's big decisions, gave me the confidence to explore the world. You would have been so proud of this book – and probably would have thought I was a little mad to take it on! Your positivity and indomitable Irish spirit carried me through life's challenges, and I hope I'm keeping that legacy alive today.

To the faculty at the University of Salford, thank you for believing in me when my confidence in my own academic abilities was at its lowest. Beyond the friendships and great memories, my time at Salford ignited my passion for learning, especially in the fields of hospitality and tourism. A special thank you to Patrick Trodden, who instilled in us the simple yet profound philosophy: 'When it comes to learning, be a sponge, not a sieve.' Those words have stayed with me, shaping my approach to both travel and life.

To my sports team coaches, thank you for opening the doors to unforgettable adventures and for giving me the opportunity to travel abroad to play football and water polo during my teenage years. Those experiences not only taught me valuable lessons about teamwork and resilience but also created memories that have lasted a lifetime. It was an incredible blend of competition, camaraderie and pure joy that shaped my love for exploration and set the foundation for so much of what I value today.

To Anthony Bourdain, your storytelling – both in writing and on screen – was a constant source of inspiration to me. Your work captured the essence of food, culture and travel, and your loss was felt deeply by so many. You left an indelible mark on the world, and your legacy continues to inspire me.

To the countless people I've met during my travels and throughout my career in youth travel: thank you for the memories, the lessons and the laughter. You've enriched my life in ways words can barely express. So many of you are still friends now and I treasure our friendship through work and travel.

To the young people out there – those who challenge the norm, step out of their comfort zones and embrace the unknown to explore the world – this book is for you. Your curiosity, resilience and courage to see the world differently inspire an entire industry. You remind us of the boundless possibilities that come from travel and cultural exchange.

As long as you have the desire to discover new horizons, this industry will be here for you – dedicated to creating opportunities that allow you to grow, connect and make your mark as a global citizen. Keep dreaming, keep exploring and keep shaping the future. The world is waiting for you.

Finally, to you, the reader: thank you for picking up this book. The fact that you've taken the time to read it is humbling and means more to me than I can say. If it inspires you to rethink your brand or destination, to attract more young travellers or to create unforgettable experiences that yield immense long-term value, then I will be over the moon. Youth travel has the power to transform lives – not just for the travellers themselves, but for the people, brands and destinations they touch along the way.

Introduction

I have been fortunate enough to be connected, one way or another, with youth travel for over three decades – both as the young person enjoying the experiences and also as a person running a company that delivers similar experiences to new generations of young people.

Every business has a story, and every story is shaped by the voices that tell it. This book isn't just a collection of ideas and strategies; it's a fusion of perspectives.

Why take the approach of interviews? Because the truth rarely fits neatly into a single perspective, especially in the realm of youth travel. It emerges through the dynamic interplay of experiences, ideas and reflections. In the ever-evolving world of youth travel, no one person holds all the answers. Success thrives on dialogue, learning from others and, most importantly, the art of listening.

Interviews allow us to step away from the abstract and into the tangible. They offer a window into the humanity behind travel businesses. Hopefully, this format also challenges the notion that business success can be distilled into a universal formula. Instead, it celebrates the diversity of approaches, the unexpected detours and the individuality of every journey. You'll find conversations here that are candid, sometimes provocative and always thought-provoking. Many of these insights are born out of a personal love for travel and a desire to pass those experiences on to younger generations.

Presenting this book as a series of interviews ensures the authenticity of the voices behind the wisdom. It's not about offering a single, definitive view but about letting multiple perspectives shine. These interviews aren't just stories or case studies; they're glimpses into real experiences, with lessons as varied and nuanced as the people who share them.

My first experience of youth travel was in 1994, when through my school I headed to the town of Offenburg, in the Black Forest in Germany. Heading off from school by coach, this was my first experience at the age of 13 of independent travel from my parents. I was housed in a German family's home with a student of a similar age with the aims of improving my German language skills. A few months later, the same student was to head to the UK and live with my family in London. Even though it was many years ago, I remember the excitement of travelling by myself for the first time and the mixture of nerves at not speaking the local language well with the desire to see new environments and meet new people. In essence, this is the core of what youth travel delivers – experiential travel allowing young people to learn new cultures.

Following on from Germany, I was very fortunate to play a competitive level of the sport water polo, and in the summer of 1997 our team, who were national champions at the time, headed to Australia and New Zealand to travel and play matches for almost four weeks. Again, independent of parents, this trip couldn't have been further from home, time or distance wise. Having only known a few people that had been that far from the UK and only seeing Australia mainly in sports matches or television shows like *Neighbours*, to being there was incredibly exciting. I still remember in vivid detail that whole month. From landing in New Zealand after 28 hours of flying and playing the New Zealand U15 team in the Commonwealth Games pool to seeing the Sydney Opera House for the first time, the trip touched on all senses. I think this was the moment that I knew I had the bug for travel and I wanted to explore the world further. It also garnered my love of food and tasting new things. These initial experiences led me to want to explore the globe more.

According to UN Tourism, young travellers are those between 15 and 29 years. However, the WYSE Travel Confederation report in 2011 says the traditional age of 18–24 years for youth has now shifted to 15–30+ years. Different countries have different practices and therefore the exact definition of what a youth traveller is, is dependent on the country that is hosting that young person.[1] Some will define it by their own visa regulations, for instance the UK and Australia have a reciprocal youth Working Holiday visa that allows people between the ages of 18 and 35 to go to each country for up to two years and do paid work.[2]

My next adventures again came through sport with school football trips to both Paris and Barcelona. For me, football being my favourite sport, having the option to play in another city was just incredible and combine it with cultural experiences in two cities I had never been to. The older I became in my teens, the more I craved these moments and vowed to take a gap year after my exams to explore. On graduation from school in 1999, I decided to combine work experience in London with some adventures abroad. After working in a hotel and a French restaurant in London for eight months, I set off for a month in Hong Kong and Thailand with my then partner. This was another sensory overload: the noises, the people and of course the food!

On my return, I met with two school friends and we embarked on an interrailing trip across Europe. With nothing but 15 kg backpacks, a book with train times and no return flight, we started in Barcelona (I had to go back!) and finished in Amsterdam. Thirty days, 13 countries, endless stories and characters met. For my friends and I, we had the adventure of a lifetime. We all had European roots and retraced some of those on the journey. My experience of seeing where the 'Lowys' lived in Teplice, Czechia, before World War II was enlightening.

After three years of studying hospitality and tourism at university, I was eager to head off again. This time I embarked on a longer journey, travelling across Southeast Asia to live and work in Sydney and come back with ideas of where I wanted my career to go. After a summer of working and saving, a friend from university and I decided to go for it. We booked around-the-world tickets, leaving London on 1 October 2003 and landing in Hanoi, Vietnam with the aim of being back in London by October 2004. Not all legs were confirmed, but we aimed to be in Sydney for the middle of December, leaving from Singapore. The rest was just detail!

This was the start of my journey with online travel, where I bought my ticket online and then had to go and collect it from the travel agent. We also booked our first night in Hanoi at a guesthouse through a website called hostelworld.com. This shows just how fast travel was starting to change; within three years my travel had gone from purchasing in a travel agent on a high street to online. Little did I know how the next three or four years would transform the whole travel sector to one of the biggest online sectors in the world. After a number of months travelling on local buses through southeast Asia, we ended up in Sydney, where I started working in youth travel for the first time (a hostel in Sydney called Wake Up!). After some months living and working in Sydney, both at a hostel and an Italian restaurant, I travelled around Australia, followed by travelling across New Zealand and islands in the South Pacific before finishing off the trip with a number of stops in the US. It was a journey of a lifetime that had a positive impact on my career in many ways.

So, why is this relevant? Well, many of the destinations I visited in those formative years have been places I have continued to visit for both work or pleasure as an adult. I have developed a special connection with these places that goes beyond the connection I would have had, if I had not travelled in my formative years.

Cambodia is a place I've gone back to twice. Once for a charity funded work-related project whereby I trained the team at Rosy's Guesthouse in Siem Reap in everything from online marketing to revenue management, while also working with three different charities. Having visited seven years prior, the opportunity was too good to turn down and it was an extraordinary two weeks. The warmth of the people was just remarkable and the fact that I could genuinely make a difference was something that meant a lot to me then, as it still does now.

The next visit was with my wife on the way back from a wedding in Indonesia. My wife had not been to that region, and when deciding on where to go, I wanted to show her not only the place where I had done the charity projects, but also the magical sites of Angkor Wat and the other temples of that region.

No advertising, no social media targeting, no travel agent directing me – this was just purely an emotional connection from prior experiences that drew me back to the country. Compared to that first visit, my visits as an adult brought back more financial value to Cambodia directly and indirectly. When I was there working, there was the indirect value brought to Cambodia with the upskilling of the team members, but also the fact that the project was shared with thousands of people through a blog. Directly, we stayed in nicer hotels with a larger budget, ate in higher value restaurants and had a more premium guided tour of the temples. All of this without any tourism board needing to push us to the destination. However, I doubt if I would have done the project, or the trip for pleasure, if I hadn't had those magical six days as part of my backpacking trip in 2003.

The famous quote that states that it costs five times as much to attract a new customer as it does to keep an existing one[3] is so applicable to the power this type of tourism brings. Rather than fight for customers at a premium price point in the luxury space, why not try to attract your customers in when they are younger, and create loyalty that lasts a lifetime. In a book written by the former CEO of Saatchi and Saatchi, Kevin Roberts, he explained the concept of *Lovemarks: The future beyond brands* – companies and therefore destinations where loyalty is built emotionally and sometime even beyond reason.[4] This is achievable for destinations and brands working in the travel sector but why do so many focus their attention of the top end of the market?

The trend over the past decade towards valuing 'experiences over possessions', as explored in *Stuffocation* by James Wallman,[5] has transformed the travel industry, especially among younger generations. Today's youth are less interested in accumulating material goods, choosing instead to invest in unique, memorable experiences. For destinations and brands, this shift emphasizes the importance of offering immersive, photo-worthy moments that resonate with this demographic. By catering to these desires, brands and destinations can build loyalty, attracting young travellers who will not only return but also advocate for these experiences through social media, thus fuelling further interest and growth.

The following chapters of this book will explore best practice in targeting this market through interviews with industry experts, as well as the challenges that surround targeting the youth market. Hopefully this book will help you build a destination or brand with loyalty for a lifetime.

Notes

1   UNWTO. Global Report on the Power of Youth Travel, WYSETC, 2016. www.wysetc.org/wp-content/uploads/2016/03/Global-Report_Power-of-Youth-Travel_2016.pdf (archived at https://perma.cc/JB77-CB28)
2   Australian Government. Working Holiday Maker (WHM) program, Australian Government, 2024. immi.homeaffairs.gov.au/what-we-do/ whm -program/latest-news (archived at https://perma.cc/P35X-CP7H)
3   A Gallo. The value of keeping the right customers, Harvard Business Review, 2014. hbr.org/2014/10/the-value-of-keeping-the-right-customers (archived at https://perma.cc/JZH9-ASPK)
4   K Roberts (2004) *Lovemarks: The future beyond brands*, Powerhouse Books, New York
5   J Wallman (2015) *Stuffocation: Living more with less*, Penguin, London

1

What is youth travel?

This is a question that is often asked and, to be honest, there is no single, straightforward answer. Youth travel, at its core, is an evolving concept, shaped by age definitions, travel purposes, cultural shifts and economic factors. While some definitions focus purely on age ranges – establishing upper and lower limits – others broaden the scope to include the reasons or purpose behind the travel. For example, an international student may be considered a 'youth traveller' – yet they could be pursuing a master's degree or a PhD at an age well over 40. This multifaceted nature has led destinations, brands and governments to tweak their interpretations of what qualifies as youth travel, each tailoring it to suit their objectives and demographics.

In the UK, the British Educational Travel Association (a trade body that represents youth, student and educational travel business) defines the market as between 11 and 35 years old.[1] That lower age range often includes school, language or even sports groups whereas the older side of the spectrum is working holidaymakers, postgraduate students or leisure travellers. Looking at the membership of the organization, you can see there is a mix of language schools and adventure camps for the younger groups and backpacker hostels, adventure travel organizations and working holiday programmes for those over 18, which gives you an idea of the scope and scale of the sector that defines itself as youth.

The evolution of youth travel

The concept of youth travel has come a long way. Some of you may remember the iconic Club 18–30, a package holiday brand that catered to young British travellers seeking sun and nightlife in destinations like Ibiza or

Magaluf. These trips became synonymous with youthful exuberance and the hedonistic side of travel. However, the perception of what constitutes youth travel has evolved significantly. Brands like Contiki which now dominate the youth-focused travel space have extended their age range from the traditional 18–30 bracket to 18–35. This shift acknowledges the changing aspirations of young travellers who are exploring for longer periods, pursuing more purposeful experiences and integrating travel into various stages of their lives.

At the lower end of the age scale, school trips often begin at 11 years old. These 'semi-independent' travel experiences – where young people travel without their parents but in a structured, supervised setting – are becoming increasingly popular. Whether for educational, cultural or sports-related purposes, such trips provide young people with their first taste of independence, often sparking a lifelong curiosity about the world. These experiences also contribute to personal development, fostering resilience, problem-solving skills and cultural awareness at a formative age.

Government policies and youth travel

A lot of these age ranges are dictated by government policy with regards to visa regulations. The 'working holiday' style visa, often known as 'youth mobility schemes', range in ages from 18 up to 35 years old but also in terms of length of stay, ranging from six months to two years. Some of these visas even stipulate where and what industries you can work in. In Australia, for example, before you can apply for your Working Holiday visa extension (whereby you stay for a second year rather than just one) you need to complete three months, or 88 days, of 'specified work' in certain industries and locations.[2] These young people can work in industries like farming, fishing and construction, and even bushfire recovery work in designated disaster areas, towards their specified work requirements. This is a win–win for governments as this brings in tourism to the country and fills gaps in the industries that need the support to keep the rest of the economy moving.

The same goes for the criteria with regards to students heading to a destination for either a university-level degree, or younger, for a language style programme.

Without a sensible approach to these visas, a destination can market all they want to young people, but if the destination appears unwelcoming or has too many hurdles to enter, then young people will go elsewhere. One example of this was the UK leaving the European Union, therefore requiring young EU citizens to not only require a visa to work but also to study, hence leading to a sharp drop in numbers of students from the EU to the UK.[3] This also happened in the US through the Trump administration, when international student numbers dropped due to political unrest between the US and China.[4]

Marketing youth travel: The Millennial and Gen Z divide

When looking at this age range from a marketing perspective, it spans two distinct groups: Millennials and Gen Z. Millennials are classified as those born between 1981 and 1996, and Gen Z are those born between 1997 and 2012. These age ranges are broad, and depending on where they are based there may be a number of generations in their family who travelled at a young age. The early Millennials are now in their 40s and, like me, would have experienced youth travel in a very different way to what Gen Z are now experiencing. What is interesting is that the boom in budget travel in Europe led by the likes of EasyJet and Ryanair in the mid-1990s onwards, combined with an increase in school travel in that timeline, means that many more parents in Europe have experienced youth travel. In turn, they have become the 'influencers' to the Gen Z audience, often encouraging their children to experience the world in a way that they did 20 or 30 years ago. This is where the long-term return is for destination and brands. Loyalty that is created from unique and life-changing youth travel experiences is passed on through generations. Although those experiences may be different to their parents, there is still that reference point and interest in following that well-trodden path.

Educational benefits of youth travel

This diversity in youth travel experiences highlights how the purpose of a trip can shape the nature of the educational benefits gained. For instance, school group trips often centre on structured learning goals, such as language acquisition, where students immerse themselves in the local language and culture to build skills directly relevant to their studies. On the other hand, around-the-world backpacking journeys and volunteering experiences introduce young travellers to different cultures, customs and worldviews, fostering personal growth through informal, immersive learning. This type of travel builds cultural awareness, adaptability and a broader understanding of the world, delivering educational value in a less structured but equally impactful way.

The post-Covid phenomenon: 'Revenge travel'

The Covid-19 pandemic brought the world to a standstill, but it also gave rise to a unique phenomenon: 'revenge travel'. After prolonged lockdowns and travel restrictions, people – particularly young people – felt an intensified desire to explore. This trend was marked by a shift towards 'travel with purpose', as individuals sought experiences that were not only enjoyable but also meaningful. Ironically, this aligns with the essence of youth travel, which has long been associated with purposeful exploration, whether through education, adventure or cultural immersion.

However, despite this alignment, youth travel is often undervalued within the broader tourism industry. It is frequently perceived as low-value and low-impact, an assumption that overlooks its far-reaching benefits. Youth travellers are not just consumers; they are ambassadors, influencers and future repeat visitors. Their early experiences with a destination or brand can foster lifelong loyalty, generating long-term value that extends well beyond their initial trip.

A broad definition of youth travel

In summary, I would define youth travel as encompassing people who are 11–35 years old, travelling independently or in a group, within their home country or abroad. Given the way the youth travel industry has evolved, purpose is less important these days with regards to the definition; as there are so many different options for young people to choose from, it is not as relevant as it was in the past.

As we go through this book, we will hear the views from sector experts from different brands, destinations and associations about how they see the industry's strengths and challenges and where others can implement strategies to enhance their youth market and create a lifelong loyalty to their brand or destination.

Notes

1 BETA. Great Britain: At the heart of global youth travel, BETA, nd. www. betauk.com/gb-heart-of-global-youth-travel/ (archived at https://perma.cc/ TC2Q-Y89G)

2 Gov.UK. Youth Mobility Scheme visa, Gov.UK, nd. www.gov.uk/youth-mobility (archived at https://perma.cc/7E2Y-ZEQY)

3 O Lisa and A Richard. Number of EU students enrolling in UK universities halves post-Brexit, *Guardian*, 27 January 2023. www.theguardian.com/ education/2023/jan/27/number-eu-students-enrolling-uk-universities-down-half-since-brexit (archived at https://perma.cc/LUW4-RBWF)

4 S Stephanie. Amid 'Trump effect' fear, 40% of colleges see dip in foreign applicants, *The New York Times*, 16 March 2017. www.nytimes.com/2017/03/16/us/ international-students-us-colleges-trump.html (archived at https://perma.cc/8CQ5-8Q8P)

2

The value of the youth travel market

This chapter takes a deep dive into the shifts and milestones that have marked the journey of youth travel, as well as the challenges that remain. Leading this conversation is Patrik Pavlacic, Chief Intelligence Officer at Bonard, a research company specializing in strategic advisory services in international education and student travel.

For more than 15 years, Patrik has been at the forefront of research in youth, student and educational travel, witnessing its growth from a niche market to one that attracts serious attention from governments, investors and institutions. The rise of digital technology and the democratization of global travel have played pivotal roles in this evolution. However, while the sector has gained momentum, it has also faced obstacles, from the logistical complexities of data collection to the lingering impacts of the Covid-19 pandemic.

As you read this chapter, you will gain insights into how destinations have adapted (or failed to adapt) to meet the needs of a growing and evolving youth travel market. You'll also find reflections on why some destinations succeed in their marketing and data strategies while others lag behind. Patrik's expertise provides a roadmap for emerging destinations seeking to enter the youth travel sector, emphasizing the importance of a comprehensive approach that incorporates housing, infrastructure and marketing.

Introducing Bonard and its role in youth travel research

Stephen: Patrik, it would be interesting for people to just understand what Bonard do as an organization and what your role is there.

Patrik: So, Bonard is an independent research company. We specialize in strategic advisory services within international education and student travel. I've represented the company for the last 15 years and in my role as the Chief Intelligence Officer, I've helped destinations develop their data collection systems and spearheaded consultancy projects that facilitate informed decision-making.

The evolution of youth and student travel over 15 years

Stephen: You have done 15 years of research into youth, student and educational travel. How has it changed over those years? You must have seen a lot of change.

Patrik: Absolutely. I recall when I first joined the company; the youth and student travel industry was relatively niche regarding government attention and investor interest. Over the last 15 years, however, that has changed dramatically. Government stakeholders, educational institutions and the entire supply chain have begun to recognize this lucrative sector. Consequently, numerous research, awareness and engagement initiatives have been launched, enhancing the industry's reputation and relevance.

It's no surprise to see a significant influx of private investment in this area. From my experience working with data, I've observed an increase in the availability of key mobility statistics, which are now published more frequently. While there are still variations from one destination to another, the sector is now better documented in terms of key performance indicators than it was 10 or 15 years ago.

However, we still have a long way to go in fully understanding the student journey, including aspects like how brand affiliation affects future decision-making and how preferences evolve with age.

The impact of technology on youth travel trends and best practices in data utilization

Stephen: Technology has evolved dramatically over the past 15 years. The days of relying on a Lonely Planet Guide book for travel information are long gone. This rapid advancement presents challenges for researchers, especially when trends shift almost overnight – like with the rise of TikTok – rendering previous statistics outdated.

Given your experience working with various destinations, governments and tourism boards, which destinations are excelling in data utilization?

Additionally, what can emerging destinations do to effectively grasp the youth and student market?

Patrik: Sure. I'd like to take a moment to reflect on the evolution of data collection in the youth travel sector. When I first began in this position, I often cited Australia as a model for effectively tracking incoming students and young travellers. Their approach to reporting frequency and depth of data was, in many ways, the gold standard.

Initially, many destinations expressed a strong desire to enhance their data collection efforts, and funding was available for this purpose. However, there was often a disconnect between data collection and its practical applications. Fast forward to the present, especially before the Covid-19 pandemic and we've seen a significant increase in destinations ramping up their information-sharing efforts. Governments have made remarkable strides in data availability, implementing dashboards that enhance transparency, foster collaboration and drive innovation.

There is a clear correlation between destinations where data collection is prioritized and their overall success. Funding initiatives and frameworks for data collection typically yield better results. Consequently, the differences between Australia and other English-speaking countries in terms of data utilization have diminished over time.

Now, we also have access to monthly statistics on visa approvals, which are crucial for student mobility. Emerging destinations like Malaysia and the Philippines are becoming more transparent about their data and reporting practices, leading to improvements in quality. This progress makes international comparisons much easier than they were 10 to 15 years ago.

If you look at the global landscape, you'll notice that various destinations are stepping up their data initiatives. The Middle East, in particular, has experienced significant growth in information availability, with stakeholders, institutions and student travel operators increasingly utilizing data. Overall, much has changed and the differences between destinations in this regard have become quite marginal.

Stephen: In summary, for destinations that are not currently attracting youth, it's crucial to implement a robust data collection system. This allows you to track your efforts effectively. For example, if you're running a web marketing campaign, you'd naturally monitor the data, yet some destinations or countries may desire to attract young people without tracking how successful those initiatives are.

This advice applies not just to destinations but also to brands looking to launch youth campaigns. With your global perspective on youth and student travel, could you share the key headline statistics for this sector?

Is it still experiencing rapid growth? Has there been any slowdown, and how has the rebound been post-Covid? It would be helpful to get a broad overview of where we stand now.

Patrik: Before Covid, the industry was booming. We saw student numbers growing at nearly double-digit rates, which significantly affected supply. New businesses specializing in international student travel emerged, alongside local suppliers. The education sector was flourishing, with new language schools and education providers being established and transnational education gaining traction as brands set up branch campuses or language schools in students' home countries.

However, the Covid-19 pandemic has had lasting effects, and certain sectors are still recovering. For example, English language travel to major English-speaking destinations remains below 80 per cent of pre-pandemic levels. In contrast, higher education has shown remarkable resilience; the number of enrolled students at higher education institutions has largely remained stable. Consequently, student housing continues to be one of the top-performing sectors.

In terms of student group mobility – particularly junior group travel – these areas experienced significant impacts. Nonetheless, we observed a strong rebound in 2022, which carried into 2023, establishing a new benchmark for various data metrics. Overall, we are now seeing recovery that aligns closely with the numbers from 2019.

Post-Covid recovery in youth and student travel

Stephen: The world reopened at different stages, but it's largely open now. I believe we have a solid baseline for comparison with 2019. However, we've observed that, with the rebound in international student numbers – especially in higher education – governments are beginning to slow down the issuance of visas and impose caps.

Do you think this trend will continue until the infrastructure can accommodate the influx of international students? Or do you believe many countries have reached their saturation point and are now hesitant to accept more students moving forward?

Patrik: I believe the sector has become a victim of its own success. For many years, we saw a scenario where demand consistently outstripped supply, particularly in terms of student housing and associated services. This has proven to be a significant issue, especially in destinations like Canada, Australia, but also the Netherlands. The rapid growth they experienced was not matched by adequate improvements in supply.

For instance, the housing situation in these countries has not been systematically addressed for over 20 years. Their popularity among students surged, without the necessary steps taken to ensure these students were properly housed. This has led to challenges, compounded by the political aspect of trying to curb net migration.

Consequently, these destinations may need a few years to rebuild their brands and restore trust as welcoming destinations. The changes in policies have affected the entire supply chain, making it crucial to establish trust between Australia and Canada as destinations and their stakeholders, including students, parents and education agents.

As soon as students realized there were barriers to entering these countries, they shifted their focus to alternative destinations. Therefore, visa policy changes are indeed reshaping student mobility trends. However, I firmly believe that Australia and Canada will recover. They need to focus on building the student housing supply, because their value proposition remains highly attractive for students. Eventually, we will reach a point of equilibrium where these destinations will reinstate themselves as open.

The speed at which these adjustments are made is vital, especially concerning housing shortages. In the UK, for example, the provision rate – the ratio of beds to students – is about 30 per cent. In contrast, Australia has cities with provision rates as low as 6 per cent or 9 per cent and Canada is above 10 per cent. The gap is significant and it will take time for investors, developers and operators to become part of the solution and elevate supply levels to an acceptable standard.

TIP

This shows the knock-on effect of youth tourism and the importance of thinking about the whole ecosystem. It is fine for a country to attract lots of young people to their universities, but if there is not enough housing for the students attending, then it causes strain far beyond the sector but into areas such as housing and facilities for locals.

Youth travel's economic contributions

Stephen: In various interviews throughout the book, we discuss the importance of a comprehensive ecosystem for youth travel. This includes accessible transportation options like trains and planes, visa processes, housing and activities. Each of these elements must be carefully considered. For instance, when cities engage in placemaking, some may face over-tourism, while others struggle to develop their tourism industry. If a city plans to establish a university targeting international students, it must ensure that all components work seamlessly together. If one part fails, it can create significant issues and hinder traction.

Take Australia, for example. It's a long-haul destination with its vast distances. If students arrive and find insufficient or unaffordable housing, or are unable to work alongside their studies, it becomes a challenging sell, despite its popularity.

As for the global state of youth and student travel, it's valued in the billions. Do we have any recent statistics regarding the current state, or are we still awaiting post-Covid updates?

Patrik: One of the most surprising aspects of this vast industry is the lack of precise figures regarding its economic impact. As a researcher, I would hesitate to provide a specific number, but we know it's in the billions. Some destinations, like the UK, Australia, the US and Canada, have begun to understand the financial contributions of international students, particularly in higher education. National industry bodies are stepping up and providing insights into the value of English language teaching.

For instance, in the US, the higher education sector is valued at over $40 billion[1] while the UK's higher education sector is just above that figure. This makes education a significant export item for governments.

While this segment is the most documented within global youth and student travel, my hope is that we achieve similar transparency for other sectors. I believe that governments need to provide the framework, funding and guidance to institutions to facilitate this transparency.

One focus of my professional career has been analysing the journeys of students and young travellers, assessing how destinations can capitalize on their experiences. However, I have found that there is very little comprehensive research done in this area. Typically, ad hoc research is conducted every five years, offering only limited insights. For example, in the UK, English UK conducted a survey in 2018 that indicated about 80 per cent of their customers were considering returning to the UK in the future.[2] While this is a useful indication, reference points like this are scarce.

When destinations conduct inbound passenger surveys, very few ask about previous visits or whether travellers plan to return after their visit. This information could be invaluable for planning digital marketing outreach and campaigns to attract repeat visitors. The economic contributions of returning travellers are substantial, and we know tourism is vital to local economies.

It's striking to see the gap in how destinations approach this data. There's a lack of understanding of the returning customer and their impact. Once we can put a value on these figures, I believe many destinations will be more open to regularly collecting this information.

Stephen: It's quite astonishing, really. The youth travel sector is truly global, yet we lack comprehensive global statistics to support it. While researching for this book, I've found it challenging due to the absence of a clear definition for youth travel; it varies significantly across contexts. There's a lot of estimated data available globally, which is quite different from more established sectors, like housing. For example, we have accurate figures on how many homes are being built worldwide, their value and the employment figures associated with that.

This disparity highlights an interesting gap and I believe it's an area that organizations like Bonard could focus on in the future.

Patrik: There's an interesting analogy to consider, though it may be somewhat outdated now. For a long time, governments and stakeholders viewed youth travellers as part of a lower tier of travel. However, based on our extensive work with leading organizations, we know this perception is incorrect. In fact, the spending of youth travellers often exceeds that of the average tourist. Students and young people typically stay longer in destinations, leading to a significantly higher economic contribution than that of an average tourist.

Thus, there has been an ongoing effort to demystify these misconceptions about youth travel.

The long-term value of youth travel

Stephen: On that note, I want to emphasize a key theme of this book: for destinations and brands to recognize that investing in youth travel yields long-term benefits. Engaging with young travellers not only brings them to your destination but also cultivates their loyalty over time. This results in repeat visits, referrals to friends, the establishment of businesses and the return of families to explore the destination.

Are you aware of any countries or travel brands that are effectively tracking these long-term impacts? If so, what statistics do they have? If not, do you have any insights into why there might be a lack of tracking in this area?

Patrik: To be completely honest, I don't have a clear answer. One of the primary reasons is the complexity involved, which requires the engagement of multiple stakeholders. Tracking returning young travellers can be challenging, especially if they don't stay at the same hotel, use the same service provider or enrol in the same educational institution.

In some destinations, there have been attempts to track these patterns, particularly from the education sector. For instance, Australia has implemented a framework where students start with a language course, progress to other providers and eventually enrol in higher education institutions. However, this concept is still in its infancy and I can't speak to the actual effectiveness of tracking mobility in this context.

Monitoring youth and student travel fundamentally revolves around those initial experiences in a foreign destination, such as summer camps or student group travel. For this tracking to be successful, there needs to be an official understanding at the destination level of how to implement these systems. The potential benefits of such initiatives would be immense, yet it seems that many people have yet to consider this, despite a widespread awareness of the value attached to it.

TIP

This highlights the long-term value of young people coming to your destination. From school age to mature learner, you can build lifelong affinity, especially through high-level experiential learning experiences.

Stephen: One of the interviews I conducted touched on how banks target university students with special offers, like student fares. In reality, it's quite rare for people to switch banks once they've set up their savings account and credit card. Similarly, when it comes to flights, travellers tend to stick with one of the major airlines, such as Star Alliance or One World. Once you get your round-the-world ticket and start accumulating points, you continue to travel with them, moving from the back of the plane to the front – at least, in my case, to somewhere in the middle!

From a destination perspective, there are numerous entry points to attract youth travellers. However, many don't fully recognize the long-term benefits of investing in this market. They might intuitively understand that a significant youth marketing campaign from 20 years ago resulted in a busy tourism sector today. Still, they often lack the data to measure this impact. For instance, the IPS surveys in the UK offer some insights, but there's a gap in understanding how past youth travellers have returned for other experiences.

Numerous individuals I've interviewed for this book have shared stories about their youth travel experiences and their continued engagement with the industry. They've revisited destinations they travelled to in their younger years. It might be worthwhile for organizations like Bonard to explore how to effectively track these individuals over time, potentially reconnecting with them 20 years later.

For those reading this wanting a push for youth travel, what trends do you think they need to be thinking about, so that initial investment is successful both short-term and long-term?

Patrik: That's an insightful question. It connects back to our earlier discussion about the importance of fostering brand loyalty from the outset. This goes beyond merely identifying a target audience; it necessitates a deep understanding of Gen Z, a demographic that is markedly different from previous generations. Brands face the challenge of adapting to this shift in consumer behaviour, moving away from a predictable environment.

To effectively engage this audience, I strongly recommend conducting thorough research to comprehend what resonates with them. This approach is vital for many providers and companies in the youth travel sector. The data clearly indicates that those who invest time in understanding their audience and communicate in a language that speaks to them enjoy a higher return on investment.

Another crucial aspect is the growing importance of sustainability and climate change. Businesses that demonstrate a commitment to these values are more likely to attract students who prioritize such considerations. Furthermore, institutions that focus on experiential programmes tend to outperform those relying solely on traditional offerings. There is a notable shift towards less conventional travel experiences.

The pandemic highlighted a surge in interest in outdoor activities, which continue to drive travel trends today. Additionally, we are witnessing a blend of travel with professional development opportunities. This trend is particularly evident in Asia, where parents seek to combine international travel with educational experiences for their children.

In this context, brands should promote trips that reflect a positive journey – where young travellers are not merely escaping their daily lives but actively pursuing enriching experiences. The pursuit of memorable, transformative moments has become a central tenet of youth and student mobility.

This evolution underscores the importance of offering experiences that are 'worth writing home about'. In a competitive market where students have a plethora of options, differentiation becomes crucial. The landscape has expanded significantly, with a vast array of providers and offerings.

Consequently, comprehensive research and data utilization are essential for maintaining a competitive edge.

TIP

If you do not have the data on the volume and value of youth travel to your destination or brand, then you really should try to obtain it. Using a professional third party adds weight to any study that is done, and is often performed via a trade association to ensure the breadth of the sector is involved. It is important to think about what you want to find out before embarking on the study, and whether you want a consumer research study or a value study interviewing companies within the sector. These studies allow industry to lobby national and local governments with accurate figures to explain what can be won and lost from various policies.

Stephen: Thank you for your insights, Patrik. It's fascinating to consider how much the landscape of youth travel has evolved over the last 15 to 20 years. The expansion of countries available to student travellers, coupled with the diverse range of universities offering innovative subjects – like data science, sports science and neuroscience – has transformed the travel experience.

In this digital age, where younger generations often find themselves glued to screens, the value of human experiences – those that engage all the senses – has never been more crucial. Youth travel uniquely positions itself to offer these immersive experiences, standing out in a crowded marketplace.

As we discuss the potential of this sector, it's clear that while we have a solid grasp of the short-term volume of visitors, understanding the long-term positive impacts on destinations and brands remains a challenge. The current focus is often on visa policies and immediate metrics, overshadowing the lasting value these young travellers bring.

Looking ahead, I share your optimism about the role of AI in shaping our understanding of global youth travel. The ability to analyse data comprehensively will undoubtedly lead to clearer insights into the true value and volume of this vibrant sector. Thank you again for sharing your knowledge; it's evident that the journey of youth travel is just beginning and I look forward to seeing how it unfolds in the coming years.

Conclusion

In this interview, Bonard's research and insights reveal that while youth and student travel are recovering post-Covid, challenges related to visa policies, housing shortages and infrastructure remain. However, there are significant opportunities for destinations and brands that invest in data-driven strategies and prioritize sustainability and experiential travel. Destinations that better understand the long-term impact of youth travel and effectively track their visitors stand to benefit from the growth of this vibrant sector.

KEY TAKEAWAYS

1 **Changes in youth and student travel over 15 years:** When Patrik started, youth and student travel were relatively niche and lacked government and investor attention. Over time, there's been a significant influx of private investment and government initiatives that have enhanced the reputation and documentation of this sector.

2 **Data utilization for destinations:** Destinations that invest in robust data collection see better long-term results. Emerging destinations must establish systems to track their marketing efforts and overall success in attracting young travellers. This involves not just tracking arrivals but also measuring engagement and retention. Youth and student travellers are often misunderstood. Governments sometimes view them as lower-tier visitors, when in fact they contribute significantly to the local economy. Unlike short-term tourists, youth travellers often stay longer and spend more, making them a high-value segment. For destinations, the long-term benefits of investing in youth travel are enormous, yet they often lack the data to measure its true economic impact.

3   **Impact of Covid-19 on the sector:** Pre-Covid, student numbers grew at near double-digit rates, leading to the expansion of international education and the establishment of new businesses catering to youth travel. However, the pandemic disrupted growth, with sectors like English language travel still recovering. Higher education proved more resilient, and 2023 marked a significant rebound in student mobility, although housing shortages in places like Australia and Canada remain significant issues.

4   **Visa policies and infrastructure challenges:** Governments are beginning to slow down the issuance of visas due to capacity challenges, such as housing shortages and infrastructure limits. For example, in Australia and Canada, there is a housing crisis, with supply far behind demand. These countries may need several years to rebuild trust with international students and address these issues. Without sufficient housing or essential services, destinations face challenges in attracting students long-term.

5   **Sustainability and experiential travel:** Sustainability and the pursuit of meaningful experiences are top priorities for young travellers. They are interested in travel that provides professional development opportunities and culturally immersive experiences. Brands and destinations that align with these values are more likely to succeed in capturing this market. The pandemic further highlighted the appeal of outdoor activities and adventure travel, which continue to be key drivers in youth travel trends.

6   **Importance of tracking long-term loyalty:** There is a gap in how destinations track returning travellers and long-term loyalty. Patrik notes that few destinations actively collect data on repeat visitors, yet this data could provide valuable insights for digital marketing and strategic planning. Despite the challenges, there is a tremendous opportunity for destinations and brands to track youth travellers and capitalize on their lifetime value.

7   **Future trends and the role of AI:** The role of AI in analysing travel data will be crucial moving forward. By leveraging advanced data analytics, destinations can better understand the long-term positive impacts of youth travel. AI will help researchers and destinations overcome the current limitations in data collection, allowing for more comprehensive insights into the true value of youth travel.

Notes

1 D Omar. High fees paid by international students help US universities balance their books, The World, 28 March 2024. theworld.org/stories/2024/03/28/high-fees-paid-international-students-help-us-universities-balance-their-books (archived at https://perma.cc/F6YL-MPJN)

2 English UK. International English students and their value to the UK, English UK, 2018. www.englishuk.com/uploads/assets/public_affairs/2018_mac/English_UK_Submission_to_MAC_International_Students_January_26_2018.pdf (archived at https://perma.cc/NH7Q-HUUH)

3

Attracting the youth travel market

In this chapter, I sit down with Sam Willan, an accomplished marketing consultant who has carved out a niche in the youth and student travel sector, to explore what sets successful strategies apart in this dynamic market. We dig into emerging trends in global air travel, the intricacies of appealing to young travellers and how certain destinations and brands have adapted their marketing strategies to cater to this audience.

This chapter forms a strategic guide for businesses seeking to tap into the powerful, ever-evolving force that is youth travel. Whether you're in hospitality, travel marketing or destination management, understanding the insights shared in this chapter will be critical for crafting campaigns that not only attract young audiences but also build lasting loyalty.

The growing power of youth travel

Stephen: Sam, you have spent most of your career, particularly over the last decade, working in student flights and youth travel. Why do you think youth travel is so important to destinations and brands?

Sam: Great question to start. There are three main reasons why this market is so lucrative and is so attractive. Firstly, it is an incredibly big addressable market. Secondly, being young travellers, there is a long lifetime value. Third, it is an incredibly resilient segment. These are three things that any business that is looking to enter a new market should be considering.

First off, the market size. While market estimates vary, there are close to 350 million youth travel trips taken every year, which equates to a value of somewhere in the region of $400 billion worth of value globally.[1] One thing to note is that this has almost doubled since 2009 as well. So, we're looking at a large, fast-growing industry which accounts for around 20–30 per cent of the entire global travel spend.

So, it is certainly not an insignificant segment!

Looking next at the long-term customer value angle and, as you say, my career has been very much on the airline side. Airlines see real value in this market because it's an opportunity to capture long-term customers very early in their travel life. These are travellers who are most likely making their first travel decisions, outside of those made (and funded) by their parents earlier in life. This presents a great opportunity for airlines to capture attention for the brand, but also to get them onto their own loyalty programmes and start fostering a consumer relationship. It allows airlines to build a connection from an early age with young travellers, which was backed up by a piece of research that about 80 per cent of young travellers who have a positive first experience with an airline will go on to develop a brand preference and will likely book again in the future. It's a long-term investment that is worth making.

Attracting young travellers is also complementary to other travel verticals and other traveller types for airlines. When you think of the traditional travel peaks – the corporate traveller who is flying Monday to Thursday from London to New York, the family leisure traveller who's travelling from Saturday to Saturday inside of the school holidays, for example – the peak periods where airlines don't really need any help in filling seats and certainly don't need to discount their seats to win business.

But we see from younger travellers that there is a willingness to travel outside of these peak periods, because they're not necessarily in standard nine-to-five jobs and not tied to the same schedules as corporate travellers and the wider leisure market would be. They're happy to travel outside of the peak periods in low and shoulder seasons, where the airlines need a little bit more help in terms of driving demand and getting bums on seats. So, they play a key role in building a sustainable and diversified customer base.

I worked with one airline who during Covid were reporting that about 80 per cent of their travellers were under the age of 30, which is incredible when you think about the travel population as a whole.

From a destination point of view, there's always been this perception that young travellers are budget travellers. And that's true from a day-to-day perspective. The real value, when you dig down, comes from the length of time that they will spend in any given destination. While, in general, they may be financially poor, they are often time-rich. This is reflected in the average time spent at a destination. For a youth traveller, this is 52 days, versus 10 for the more mainstream leisure traveller.

Overall, what we see is that they spend a lot more money in destination than the average holidaymaker would – around 2.5–3 times the average spends. What's more, most of that budget is spent on the ground in-destination. They're not looking to necessarily splurge on the 'travel' element – getting to and from a destination – they are prioritizing spending when they're there, which makes a huge contribution to the local economy. So the economic impact is felt much more at a local level.

They're also looking for unique experiences, which really helps from a tourism and destination point of view, who often prioritize regional dispersal, encouraging travellers to get beyond the main economic centres. Where a more mainstream traveller might spend 7 to 10 days on a beach, firmly concentrated in the tourism hotspots, these travellers want to get off the beaten track and find something different and unique. Again, as with the airlines, this flows into creating a much more sustainable and diversified customer base and revenue stream.

Investment into acquiring youth travellers will likely attract others to the destination, too. Friends will come and visit while they're there – this is particularly true of those who spend an extended period away, such as the traditional gap year or working holiday. Young career professionals will come out and join them and they will be visited by family. This has a cyclical effect of helping spread word about a destination, and they will continue to be vocal advocates for that destination long after they have left. Well over half of backpackers who have a positive experience in a destination when they're younger go back there in the future with their families as a more mainstream leisure traveller.[2]

The word of mouth, advocation and propensity to attract others to the destination is a huge opportunity for building soft power. It's something that we generally don't talk enough about – the immense benefit that can be driven by word of mouth and the positive image on the world stage that these youth travellers will bring. It really helps build the global image of a destination on the world stage.

It's clear that we have been through a tumultuous time in the last 10–20 years. Plenty of change, plenty of disruption – but this is a dynamic and adventurous group of people. They're often the last to stop travelling when things get crazy and, often, are the first to start travelling again, once things start to return to normal. We certainly saw this most pronounced with Covid, but we see it time and time again. In 2017, when London Bridge was subject to a horrendous terror attack, we looked at the desire to travel or continue to travel in the face of these issues and the propensity and desire to travel went up. This is a very defiant generation, a generation who have grown up with this uncertainty and a disrupted world. They simply won't compromise on these travel experiences, no matter what the world throws at them.

I mentioned earlier in this chapter that one of the major airlines out of the US through Covid was reporting at one point that 80 per cent of their travellers across all their flights were under 30. That speaks volumes to who was and wasn't travelling during that time. As any business, you're always looking to rid yourself of any uncertainty and this segment has proven themselves time and again to be incredibly resilient.

There is also increasing evidence to show that young people will travel more in a poor economy. They tend to be a little bit more insulated from the wider economic market because they haven't necessarily entered into the workforce just yet, or continue to have financial support from families. Couple this with the fact that they're looking for unique experiences and have a very short window in which to do it, and it is something that they simply won't compromise on.

If we look at other segments of the market again – the corporate meeting in New York gets cancelled, it's done and dusted. That isn't coming back. The mainstream leisure traveller who wants a week in the Maldives, for example, that holiday could be pushed to next year if needs be. But the young traveller who is looking for a once-in-a-lifetime, formative experience is going to prioritize it above anything else and will go to great lengths to ensure that their trip happens. Their heart is set on it.

TIP

This highlights the long-term value in investing in youth across the whole tourism ecosystem, from flights to accommodation, from urban to rural areas.

Stephen: There are a lot of emotive KPIs there! If you are a business as opposed to a destination, you might be looking at this and thinking that you should really focus on this market. But there are very few really doing it well, including airlines and including a lot of other travel brands.

Sam: Absolutely – it sounds so simple once you lay out the opportunity. But yes, very few are getting it right.

Stephen: I think so, too. In different chapters of this book, we are exploring different avenues or different issues and opportunities that present themselves. Sometimes people see it as just an issue faced by large brands, but we need to think about a destination as a business. If 35 per cent of your visitors are young, then you should advertise to them as much as the luxury traveller or any other key segment. Given the understanding of their length of stay, the dispersal of revenue and propensity to become repeat visitors, it is just so important to make better, more informed decisions in terms of measuring this long-term value that young people bring.

On that topic, you've worked with many destinations and airlines. Clearly, some do it better than others. Who would you say is getting it right in this space?

Resilience in action: Lessons from Covid and beyond

Sam: The important thing to start with is that you can't rely solely on a marketing campaign. This may sound strange coming from someone who's worked in marketing his entire life, but the core fundamentals have got to be there. If you think about a good brand or a good marketing message, it is always an accelerant of a good core customer proposition. That's not something that can just be done by marketing. For a destination, this needs to start with government policy, and for an airline, or any brand for that matter, it needs to be at the heart of business and commercial decision-making.

A great example would be Tourism Australia. Attracting the youth traveller to Australia is right at the heart of their tourism decisions. They don't see this as a 'nice to have' but as something that is critical to their long-term strategy of attracting young travellers, with the knowledge that they will have a net positive impact on the economy, will attract friends and family to the country and are likely to come back in the future, delivering long-term value.

This strategy starts firmly with government policy and the Working Holiday Maker scheme, a bilateral agreement with not only the UK but over 40 other countries as well. This gives young travellers up to the age of 35, recently raised from 30, the opportunity to work and spend some time at leisure in the country over an extended period of time. What they've recognized is that, aside from the long-term benefits we have talked so much about, there is an immediate impact on supporting the economy.

This has been especially noticeable in hospitality, where we saw a huge human resource gap during Covid. For whatever reason, young Brits don't want to take up hospitality jobs, bar jobs, waiting jobs in the UK, but they will more than happily travel to Australia and do those same jobs. They earn money in the country and they will then go on to spend in the country. They just pump it back in and they bring benefits to the country because of that. Last year over a quarter of a million hospitality jobs in Australia were filled by this Working Holiday Maker scheme.

As I said at the start, Tourism Australia are savvy because they have taken the long view on this and made it a part of their core to any marketing acquisition strategy. They accept that they won't just see the value today, but also in the future, and that is something that a lot of businesses get wrong. So many businesses define their customer acquisition success as whether a customer generates more revenue in the short term than it costs to acquire them – if not, then it's not a worthwhile investment. However, by recognizing that this can be a 20, 30, 40 year cycle they are benefitting from both the short- and the long-term economic benefits.

In addition to the economic benefit, a word-of-mouth halo effect, the soft power that it builds in the country, has also been recognized. Of course, they also have a phenomenal marketing campaign to wrap it in. They always invest very heavily in their core markets, especially in the UK and Germany. There has been a lot written about Tourism Australia's marketing strategy – they really get it right. But, of course, it starts with the core fundamental proposition and a recognition of the long-term value of the investment.

TIP

As featured across all chapters, this highlights how Australia gets it right in terms of marketing to youth. It is not a one-off but a consistent belief that investing in youth brings back short- and long-term value to the country.

Stephen: With Australia being such a long-haul destination, in many ways they can't get lazy in terms of attracting people like more well-known destinations such as the US or Europe do because they've got such a large market so close by. Australia is far from everyone and they certainly don't have a huge population internally. So, they've realized that if you can invest in youth, it just makes a better destination for everyone.

Sam: Absolutely. It was well documented that they were one of the most locked-down countries during Covid. They were one of the last to reopen, but they've rebounded incredibly quickly. The 'brand' sentiment still is there for Australia, and if they had only taken a short-term view it might have been a country that could have easily been forgotten. Of course, there was an impact on tourism for the lockdown period, but they spent so long investing in these markets and building up that brand equity or mental availability of the destination so that people were ready to go back there – young travellers were ready to go back there when they were able to do so again.

Stephen: Maybe it's an Australian one, but what campaign have you either worked on or seen that has impressed you the most and do you know the impact that these campaigns have had?

Marketing to Gen Z: Authenticity, humour and digital first

Sam: One great example from the airline world is Ryanair. This is not necessarily a campaign per se, but more part of their ongoing content strategy. The one thing that Ryanair do phenomenally well is they know exactly who they are and who they aren't. Rory Sutherland, chairman at Ogilvy, said that the opposite of a good idea can also be a good idea,[3] and I think, in business, the common wisdom would be to say you have to shout about how good you are – how good customer service is, how good on-board comfort is – and never down-play yourself. Why would you? Why would you say something that on the face of it would damage your brand?

But Ryanair have a phenomenal approach to this, particularly on social, which isn't overtly targeting young travellers, but really leans into their desire for authenticity and humour. They call out the fact that you won't get a window seat, that you won't be seated with friends unless you pay. They even specifically called out one customer comment which said that Ryanair offers no benefits by responding with a meme of a flashing toilet, saying they offer 'free use of a toilet on board'. They have really recognized the intersection of humour and brand building on TikTok that builds a positive sentiment among their prospective customers. It's a smart long-term play because you're not setting anyone up for disappointment. They are being incredibly clear in who they are and who they are not, which is a critical part of your marketing positioning. They are attracting young, price conscious travellers who are willing to accept a compromise. They have shown that they really understand their audience, and this is a group of people who can detect nonsense from miles away. We're way beyond the *Mad Men* era where you can dress anything up with clever marketing, but because of how honest they are, and how they engage and entertain their audience by being self-deprecating, they're building a much more engaged and amenable audience.

What's more, it is demonstrably working – they have over one and a half million followers on TikTok consistently generating hundreds of thousands of views on all their videos, and that is far from vanity metrics. These numbers are something that can't be underestimated, especially now, when you look at the modern full-funnel marketed approach – more young travellers now are starting their search journey on TikTok and on social media than are on Google.

This is something that has been a huge acceleration in the last 12+ months. We still talk about Google as the biggest search engine but there is certainly a shift away from this being the case – especially in this demographic. Young travellers aren't looking for what we would consider the traditional user and are instead turning to TikTok and Instagram for this initial research. From a brand point of view, being on these platforms when they are initially searching and then using that to drive the rest of your marketing is critical.

Another great example shows the value of partnerships in the youth travel space. I think this is something we are seeing an increasing amount of in the market. This was a campaign between Student Universe, a student flights online travel agency working in partnership with Amazon Prime Student out of the US. It started out as a PR-led campaign that helped target and solve a very specific problem for students – that travel is incredibly expensive, and even more so over the holiday period (Hanukkah, Christmas and New Year). We worked with Amazon and found that over 76 per cent of US college students weren't going to see friends and family over the holiday period because of the cost, and it's a cost that is high through no fault of their own – simply supply and demand, it's peak travel season. We came up with the idea of the $25 'Home for the Holidays' ticket.[4] It was a way to give back into the student community and help students to get home and see friends and family for a fraction of the cost that they would otherwise have had to pay. So that in itself came from a really great place. But it really helped that brand positioning, you know, by tapping into Amazon as the most consumer-centric company on the planet. Everyone knows Amazon and what they are about. From a Student Universe point of view, where I was working at the time, it really helped drive awareness of the brand, push brand consideration over 80 per cent and earned over $16 million in earned media value completely organically.

This just shows that, with the right partner positioning, at the right time, when you are being truly authentic by specifically solving a problem for students (or young travellers), it's something that will get real buy-in. That was a fun one to work on!

TIP

You don't have to create a youth marketing campaign on your own – partner with the right brand and you can both benefit.

Stephen: I know in the past you had a guy in a bathtub of baked beans, so obviously had some fun there, too. But seriously, in terms of companies or people in influential positions reading this, it's critical to ensure that you make sure that your own brand matches with the partner brand that you are working with. The Gen Z generation are quite sensitive in terms of brands, their core values, their sustainability credentials and so forth, even though arguably travel isn't as sustainable as it could be right now. But there is a lot of investment in that space and you know there will be in the future. In reality, they want to be associated with brands that they feel uphold their social beliefs. I suppose that's how it's changed a bit over the last 10 or 20 years.

Sam: Young audiences are quick to highlight when brands fail to align with their stated social values. It's a prime example of how crucial it is to get these aspects right. However, it's also an area where mistakes can be made easily and in today's digital age those missteps can spread quickly. Once an error is made, there's virtually no place to hide.

Stephen: Definitely. You worked for a large travel brand that had multiple brands, covering huge amounts of different niches. Could you see that those young people who bought flights at Student Universe then become brand loyal for the rest of their purchasing career or their lifecycle? Was that quite a hard thing to track? Or was that even something that was valued within the organization?

Creating a strategy that lasts:
Aligning marketing, product and policy

Sam: The value that is there to the travel company would be exactly the same as we talked about earlier. Having a youth acquisition strategy as part of a wider travel group works for exactly these reasons – you have an opportunity to capture the youth and student traveller very early in their travel journey. Possibly at their first or second real travel-making decision or when they are really starting to make their own significant purchasing decisions and spending their own money. Although, in many cases, in fact more often than not, first time around they are still spending their parents' money!

The benefit from a marketing point of view, and certainly from a business point of view, of having a brand within a wider portfolio of companies means that you don't have to be all things to all people. So, it really helps to nail the brand positioning. There is less pressure to attract more customers to drive the volume, because you have another business elsewhere in the group who you know is better positioned to do that. Better positioned to meet the needs of those customers. That was certainly a benefit for me running a student brand – it allowed us to have tight targeting and positioning. It also creates an environment where it's OK to lose a customer to another brand, because the group has a brand better placed to meet their needs – and overall, as a group, you still win.

One thing I would say that is always challenging is gaining business alignment. It is a perfect example of something that works on paper or in a slide deck, but isn't always achievable in the real world.

You can picture the customer pipeline where the student and youth brand is the 'acquisition brand', and there is a nice, clean, perfect time when you make the handoff to the brand that would better serve their leisure needs, or their corporate travel needs. The reality is that, while you can track some of that and get an overall picture of the value that this is driving and start to do some cross pollination of brands, etc, one thing that is incredibly tricky to get right is that people are people! They don't have a linear journey even though brands like to think they have. So, while having this nurtured pipeline and a 'handoff' point works well on paper, the reality is that by having better brand collaboration within the group customers will make the move at the right time for them. An example of being outward facing to the customers and not just doing what suits your own business needs.

TIP

Creating a youth acquisition strategy that is bought into across the senior team of your organization provides the best platform for success in the results achieved.

Stephen: I guess if you think destination-wise, the way that a lot of brands are set up is to have this youth marketing strategist or acquisition manager focused on youth, because then you are able to track that through the lifetime journey.

So, just to wrap up, what sort of advice would you give to someone reading this where they have realized that they really do need to do something around our youth strategy, our youth marketing acquisition of how to plan it out, how to build a great strategy and how to push out a great marketing campaign?

Sam: Firstly, going back to something we talked about earlier, the core product needs to be up to scratch. We are past the days where you can come up with a clever tag line or a claim that can mask a bad core product market. With this demographic, there's always someone else they can go to other than you. So, it is imperative that you use your marketing to accelerate and amplify your core product, rather than thinking it can be a means to grow in a silo. This starts with the fundamentals of just understanding your customers' wants and needs. It sounds obvious, but it's absolutely critical to getting things right – especially in the youth market. If you're not solving a problem for your customer, then what are you doing? We already talked in depth about the Australian Working Holiday Maker scheme, but Virgin Atlantic are another great example here of a business who is just getting it right.

Stephen: So, with the Australian Working Holiday visa, if they did a great job of promoting it and attracting young travellers, but then didn't have the jobs or the hostels or the student discounted flights to get there, then the whole thing would fall apart. It really shows the need to be hyper-connected across the whole customer journey. At the super-luxury end of travel, if you've got an amazing boutique hotel in the middle of nowhere with a Michelin star restaurant, the price sensitivity is lower and people will just spend whatever to get there, because when they get there it is still so expensive. Whereas a youth traveller, because they are a bit more budget conscious, will want to make sure they get bang for their buck.

Sam: It is important to understand the different priorities within that travel ecosystem and the interplay between different products. As an airline, it's price first. They generally want to get to their destination as cheaply as possible. If it takes 48 hours to fly to Australia and they stop over twice in two different cities, that's fine because they're time-rich. However, they want to have that money to spend on experiences in destination. So, like you rightly say, it's all about finding the balance within the whole traveller journey. What is true for flights may not be true for their choice of accommodation.

It really demonstrates how, from a marketing point of view, you must be completely in tune with what your core product is and how it is specifically related to the wants and needs of your customers.

Marketing teams should be at the sharp edge of what the customer wants and should be feeding back into commercial, leadership and sales – exactly what the business needs to do to keep up. It's critical that your core product and business strategy are in lockstep with your marketing and your brand, otherwise you are only setting people up for disappointment.

It is probably the most over-used term in marketing, but being authentic to what your customer (or prospective customer) wants is the key to getting things right. Young Ryanair customers don't want polished advertising in this market. They want real and honest. They want to feel like they're talking to a friend.

Another area I have seen a lot of success with is leveraging your own customers to let them tell your brand story, co-creating your content with your customer. There has been a real acceleration of this in the youth space over the last few years. It's one of the more difficult things for marketing teams to accept and let go of the reins in your role as a brand guardian, but I have seen some great success stories.

Effectively, it is letting your customers tell your own brand story, the way they see you. This generally tends to be social first, but it can be leveraged across all marketing channels. The benefits of that are twofold. Firstly, they promote your business in their own words in a way that comes across as natural to their peer group, which is authentic. Secondly, it helps you uncover some potential pain points and things that you need to address once you start seeing this come through in the form of qualitative research. Another reason why marketing teams need to be as close to the business strategy as they are to the customer.

I would also say it is critical not to neglect investment in building your brand, especially in travel. It is a long cycle to purchase. People aren't looking to fly every week. People aren't booking a holiday every week – so there are only going to be a very small percentage of people who are in the market to purchase at any given time.

In recent years, we have become obsessed as marketers with performance marketing (e.g. paid search). It's a very simple formula – if the money in is more than the money out, then we're doing a good job. It's easy to convey its value to finance teams; however, while these tactics certainly have their role to play in capturing demand, don't neglect investing in your brand for the long term. This builds awareness and trust of your core product in the minds of the 99.9 per cent of consumers who aren't looking to purchase today. It helps position you top of mind for the point at which they are looking to buy.

This is also a differentiator from older generations as well. I saw a bit of research from Meta (Facebook) that younger generations are six times less likely to click on an ad to go through to a website than those who are aged 50 and over. So the effectiveness of the performance marketing model is starting to become less certain because you can't necessarily see that direct attribution through clicks. Younger travellers are much more likely to go directly or search elsewhere, rather than click on an ad. That's where investing in building your brand can give a real competitive advantage.

Stephen: Particularly with youth marketing, youth travel focused destinations or brands that are trying to attract those young people, you must stay relevant. Your marketing techniques have to either be on the curve or ahead of the curve if you can because things can change very quickly. When I did my round-the-world trip, I went to a shop and bought my ticket. Then by the time we came back from travelling, everyone was using Google, but when I left to travel, everyone was using Yahoo. And that's only a year! Then, as you mentioned, in a year TikTok became one of the number one search platform for travel brands. If you're not there, you're not relevant. You won't be talking to the traveller and you won't be seeing the booking. That must be a huge challenge for many marketers who have long-term plans that may need to be torn up after 6 to 12 months because they're no longer relevant.

Sam: That really speaks to the need to have a solid strategy behind your marketing outputs, because I think if you need to change your strategy every three or six months you're probably doing something wrong! But the need to change your marketing tactics, the ways in which you execute your strategy – absolutely!

That's a prime example, of why strategy is so important. Really understanding where you are playing, understanding what customers' wants and needs are, who you are targeting, how you position yourself in a market – these things should be fundamental to your business and brand plans. But you're completely right – you might need to adapt quickly on the ways in which you reach those customers. But I still think having that core positioning is critical.

Stephen: That was interesting. I think there's a lot to take away, not least the importance of being authentic and relevant in communicating with the younger generation. Some people clearly already do it much better, but there is a huge long-term opportunity for those destinations and brands looking to tap into the value of this cohort of travellers.

Conclusion

The interview with Sam, who has extensive experience in youth travel, highlights the importance of this demographic for destinations and brands. The discussion focuses on why the youth travel segment is lucrative, resilient and valuable for long-term growth. Sam explains that the youth travel market is growing fast, accounting for a significant portion of global travel spend, with travellers in this segment providing high lifetime value. Sam emphasizes the need for brands to create a long-term strategy for attracting youth travellers, using authentic marketing aligned with their values and leveraging digital platforms effectively.

KEY TAKEAWAYS

1 **Youth travel market size:** There are nearly 350 million youth travel trips per year, valued at around $400 billion globally. This has doubled since 2009.

2 **Lifetime customer value:** Youth travel offers an opportunity to capture customers early in their travel lives, fostering long-term brand loyalty.

3 **Flexibility and resilience:** Young travellers are less constrained by traditional schedules, providing travel demand in low seasons. They're also resilient, being the first to return to travel post-crisis, as demonstrated during Covid.

4 **Economic impact on destinations:** Youth travellers stay longer (52 days on average) and spend more locally than mainstream tourists, making them a key driver of local economies.

5 **Advocates and word of mouth:** Youth travellers often share their experiences, attracting friends, family and future visits. They also become loyal repeat visitors, boosting destination advocacy and soft power.

6 **Adapting to new marketing trends:** The conversation underscores the need for destinations and brands to be aware of where young travellers are searching for information, with platforms like TikTok and Instagram becoming critical.

Notes

1 Wyse Travel Confederation. The power of youth travel, Wyse Travel Confederation, nd. www.wysetc.org/research/the-power-of-youth-travel (archived at https://perma.cc/3WBM-H78T)

2 Wyse Travel Confederation. The power of youth travel, Wyse Travel Confederation, nd. www.wysetc.org/research/the-power-of-youth-travel (archived at https://perma.cc/QA9G-88ZF)

3 R Sutherland. Quotes, Good Reads, 2022. www.goodreads.com/quotes/11064798-rory-s-rules-of-alchemy-the-opposite-of-a-good-idea (archived at https://perma.cc/78FA-838P)

4 Amazon. Prime members enrolled in the young adult plan can book $25 flights for the holidays – here's how to get the limited-time deal, Amazon, 2024. www.aboutamazon.com/news/retail/amazon-25-dollar-flights-travel (archived at https://perma.cc/EC93-QHKC)

4

Building community

Hostels as key players in youth tourism

In today's rapidly evolving travel landscape, understanding the preferences and needs of youth travellers is crucial for both brands and destinations. This chapter delves into the insights shared by Kash Battacharya, the man behind the BudgetTraveller blog, whose journey over the past 15 years has seen him witness the amazing power of hostels in the youth travel sector.

My conversation with Kash reveals how hostels have transitioned from being merely budget accommodation to vibrant social hubs that attract young travellers seeking unique experiences. He emphasizes that today's youth are not just looking for a bed; they want an environment that reflects their lifestyles, interests and values. By fostering connections, facilitating cultural immersion and providing affordable yet memorable experiences, hostels have established themselves as central players in the youth travel ecosystem.

As we explore Kash's journey we will uncover the importance of recognizing hostels as integral to a destination's tourism infrastructure. We will also examine how modern marketing strategies, particularly through social media, are reshaping the way youth travellers engage with the world. This chapter not only aims to provide valuable insights into the hostel experience but also seeks to inspire stakeholders to invest in the vibrant youth travel market and harness its potential for sustainable growth.

From backpacker budget to luxury experiences

Stephen: Could you explain a bit about your blog? Additionally, I'd like to know how your focus has shifted towards video content and how long you've been involved with hostels and travelling in general.

Kash: Yes, so it's now coming to my 15th year of blogging, so I'm now one of the OGs in the room, so to speak. My blog is called the BudgetTraveller, which is all about how to travel in style on a budget. But when I started the blog 15 years ago, it was basically just about my backpacking adventures staying in hostels, because when I was a poor student coming out of university with a lot of debt I couldn't afford to stay in fancy places or hotels. The only places I could afford were hostels and, thanks to Ryanair's two pence return flights, I managed to see a lot of Europe. Hostels were the places where I could sleep, but also, I could meet other young people and expand my horizons and learn about the world. It started out as a hobby; it wasn't meant to become a full-time job. Here we are, 15 years later: I've written four books about the hostel industry, including *The Grand Hostels: Luxury hostels of the world*, which is a book all about this amazing world of what I call luxury, but not luxury in the traditional sense. Instead, it's luxury in terms of experiences, which is what today's youth traveller is looking for.

So, documenting the rise in how hostels became cool and became a place where young people feel at home and connected with others is a key focus. The blog also offers basic tips on how to save money and where to find good places to eat. Food is something that people love when travelling, just like you do and I do.

As the digital world has evolved, particularly across social media, I've shifted focus from the blog to social media. Nowadays, my partner and I create lots of cool, epic videos about destinations and places that young people might want to visit, which has become another hallmark of my work.

TIP

Are you still using just a written blog for your marketing content? Think about photo, video and integrating content into a wider social media engagement strategy. Pure written text won't give you the desired results when targeting the youth market.

The transformation of hostels

Stephen: We have known each other since you started, and I guess the whole world of these travellers has changed a lot. Obviously, your core focus was hostels. How have you seen the sector change over the last 15 years – both from a traveller perspective (as in what travellers want), and also in the different types of hostels that are now offered to them?

Kash: It's a massive change that we've been privileged to have seen. Fifteen years ago, when I started blogging, the hostel scene was beginning to change. If you go back 20 years, when I started travelling, hostels were just a clean, comfortable place to sleep, somewhere cheap and central. You'd get your bed and, if you were lucky, you'd have a bar and a cheap breakfast. That was pretty much it; there weren't many bells and whistles.

Hostels were not as flashy as they are nowadays. They were just places where people could sleep. If there was a common room, that was where you would hang out. My first hostel was the Yoho in Salzburg, which was a staple of the backpacking circuit. They had a little bar and I remember having a great time the first night and meeting people from all over the world, which made me fall in love with travel.

Australian writer Peter Moore explained that hostels can be viewed as a gateway drug to travel. That was certainly true for me; they opened up a whole world of travel.

Over time, hostels have transformed from simply being a place to sleep into vibrant social hubs where people want to spend time. Previously, you would just drop your bag and head out to explore the city. Nowadays, hostels have evolved into a new form of accommodation that people struggle to define. You'll find bars, coworking spaces, gyms and restaurants.

Today's youth travellers are looking for more than just a bed; they want an environment that supports their lifestyle. Many young people are starting their own businesses and working remotely, so they seek places where they can work, socialize and enjoy affordable drinks at the end of the night. While the desire to party remains, today's youth travellers also crave more sophisticated experiences. They want to do yoga on the rooftop or go on a hot air balloon ride.

It's encouraging to see that today's youth travellers have evolved and matured, becoming significantly more health-conscious than previous generations. I'm intrigued by how much today's travellers have changed in terms of their preferences and what they seek in a hostel experience. The dynamics are quite complex.

Changing needs of today's youth travellers

Stephen: It's fascinating to explore the various aspects we've discussed in this book, particularly the importance of travel experiences. These experiences can encompass a wide range of activities – whether one is travelling for education or attending a language school, they all contribute to the journey.

In today's world, learning has evolved significantly. For instance, travellers can now learn a language directly on their phones, a capability that simply wasn't available 20 years ago. As both travellers and travel products have evolved, it's interesting to note that not all destinations have kept pace with these changes. Given your extensive travels across Europe and around the globe, are there specific destinations that offer significantly better experiences for young travellers compared to others? What do you believe drives these differences?

Kash: I think the most important thing that youth travellers of today want is accessibility. They want to be in contact with the culture of the place. They want to meet people from that place. They don't want to be cocooned in a coach or in a fancy room full of people, drinking with others from their home country, which they could have done in London, for example. They want to go out and meet locals.

Thailand, I think, is a great example of a destination that is particularly appealing to young travellers because it's affordable and accessible. It's safe and secure. The transportation system is brilliant in cities and people speak very good English. They're very polite and welcoming. Safety is another major concern for today's youth travellers and I believe Thailand checks all the boxes. It also boasts some of the best beaches in the world, along with fantastic hostels.

It's remarkable how much the hostel sector has evolved and grown in Asia. It's one of my biggest discoveries, and it's astonishing how quickly it's changed in the past 10 years. When I walk into hostels in Asia, I often find that they are ahead of their European counterparts in terms of hospitality and service – something not many people would have said a few years ago.

Youth-centric destinations: The standouts

Stephen: When I travelled in Thailand, I found it particularly fascinating because I had explored much of Southeast Asia, with the exception of Singapore, which felt the most modernized at that point in 2003. So, looking back 21 years, it's clear that accessibility to diverse experiences has improved tremendously.

For me, the exploration of food was especially easy and I found the street food super-accessible and of course tasty. The warm welcome from locals was also a highlight. Of course, as a tourist, you had to remain vigilant to avoid being taken advantage of, but overall it was an amazing experience.

One of the themes we discussed is that, in the past, youth travel was often defined by a singular purpose. Nowadays, that definition is much broader; travel could be centred around various purposes, such as embarking on a food adventure or learning yoga across different continents.

Kash: Young travellers today prioritize accessibility and seek immersive experiences that connect them with local culture. They want to engage with the people of the places they visit rather than remain isolated in a tourist bubble. Activities like bungee jumping, hot air ballooning and paragliding are highly sought after, and these experiences are often very affordable in Asia, particularly in destinations like Thailand, as well as in Mexico and South Africa.

In contrast to other destinations, especially in Europe, Asia offers young travellers access to the experiences they've always dreamed of. It's truly magical; it opens countless horizons for exploration. The affordability of these experiences is a significant win. If you can provide both exceptional experiences and affordability, you create the ideal formula for today's youth traveller.

TIP

Before embarking on a destination campaign aimed at youth you must ask the question, 'Do we have the infrastructure to make the young traveller feel comfortable?' If not, then investing in that is vital for long-term success with this target market.

Stephen: I don't expect you to name specific destinations, but where do you think destinations and brands typically go wrong? Is it related to the product, the level of hospitality, accessibility or perhaps a combination of all these factors?

Kash: Destinations often misunderstand the value that today's youth travellers bring. We've discussed this many times, and there's often a fixation on older luxury travellers because they are perceived as high-value. However, today's youth travellers have a significant disposable income, as shown in various studies, and they also prioritize sustainability.

Youth travellers tend to spend longer in a destination, fully immersing themselves in the local culture. A critical and often-overlooked aspect of youth travel – and the backpacking hostel segment – is that the money spent goes directly into the local economy. For instance, when participating in a food tour with a local guide, travellers are putting money straight into the community. Independent hostels are deeply rooted in their neighbourhoods, serving local residents and businesses. This means there is no wastage of funds being funnelled into corporate hospitality chains. The money flows directly into the hands and pockets of locals, making youth travel not only sustainable but also highly localized. My biggest challenge has been convincing destinations of the transformative power of youth travel – not just for short-term benefits but for long-term growth as well. Unfortunately, few destinations invest in or create campaigns targeting youth travellers.

Australia is a notable example with its Working Holiday visa, while Germany ran a successful campaign in 2012, declaring it the year of youth travel. I wish more destinations would celebrate and actively seek to attract youth travellers.

Stephen: The Australians have emerged from various prior discussions as a prime example of understanding not only the short-term value but also the long-term impact of local community spending. When we consider the local economic contributions, it becomes evident that youth travellers can significantly influence destinations.

Take my experience in Koh Lanta, Thailand, for instance. That one experience was so profound that no one needs to advertise it to me – I know that if I return to Thailand, I'm going to Koh Lanta.

This understanding makes marketing much more cost-effective. Over the lifetime value of a marketing campaign, the benefits it brings can be substantial relative to the initial investment. It's fascinating how many people overlook this potential. Additionally, it's worth noting that much of the marketing aimed at youth travellers today is digital. This approach is highly trackable, allowing for a clearer assessment of return on investment, compared to traditional marketing methods such as brochures targeting luxury travellers who might spend £2,000 a night in international hotel suites.

TIP

Make sure you or your team are looking at the long-term value of your investment into marketing campaigns. If you are targeting youth, remember that creating engaging social content is better than a billboard, and although the initial return on investment may not seem like high value, the long-term value could be great. Think about using not just transaction value, but the social media related impressions and interaction with regards to spreading the word about your brand or destination.

The importance of infrastructure for youth travel

Kash: I'll share an interesting example from my conversation with the tourism board of Gelsenkirchen, a city many readers may recognize as the base for the England football team during the European Football Championships. Gelsenkirchen is a small industrial town, not known for much beyond being home to Schalke Football Club, a major football team in Germany.

The tourism board was exploring ways to attract more visitors, especially since the stadium hosts numerous large concerts. They expressed a strong interest in building a hostel. Why a hostel? Because it would help draw a younger demographic to their destination.

This approach highlights a broader perspective: if we want to change the perception of our destination, we need to create spaces where young people can gather and stay. Many destinations overlook the vital role that hostels play in introducing youth travellers to their area. By building and supporting these hostels, you're effectively creating your future visitors. These young travellers will return and, eventually, may choose to stay in different types of accommodation in the future. This example illustrates how even small destinations can strive to create the right ecosystem to attract youth travellers.

Opportunities for smaller destinations

Stephen: It's interesting to note that, when we refer to a destination, it doesn't necessarily mean an entire country. As I've observed in your work, sometimes it's very small regions that invest in promoting either the hostel experience or the various travel opportunities available. It doesn't always have to be about the country as a whole; it could just as easily focus on a specific region or even a town.

Kash: A recent memorable experience took place in a small town called Chur, Switzerland, which is one of the oldest inhabited towns from the Roman period. Located just an hour from Zurich, Chur is a beautiful town that serves as a great base for numerous hikes and adventure sports activities. A friend of mine recently converted an old jail into a hostel there – it is called Bogentrakt – placing Chur on the backpacking map of Europe and Switzerland. Previously, it was just a stopover for travellers heading to the scenic train journeys in Switzerland, like the Glacier Express.

However, now young people are coming and staying for two or three nights because of the great new hostel. The previous hostels were not up to par and this new offering is attracting a new type of visitor to the destination. The tourism board has also started supporting Bogentrakt and incorporating it into their promotional activities, which is a wise strategy. The owner of the hostel even sits on the marketing board for tourism, demonstrating smart decision-making for the future.

Unfortunately, many tourism boards miss the opportunity to involve hostel managers in planning their strategies, which could enhance their efforts to attract youth travellers.

Future trends in youth travel

Stephen: This discussion has been fascinating, especially with the examples you've shared. As we've said, a destination doesn't always have to be an entire country. A small town can also create a comprehensive youth travel strategy that draws visitors in like a magnet.

I'm curious about your thoughts on the future. How do you see the industry evolving over the next 5 to 10 years? Are there trends you've observed in Asia or Australia that might influence Europe, or vice versa, in terms of how young people travel and what types of accommodation or experiences they seek?

Kash: One area where I see significant potential for youth travel is in addressing the problem of over-tourism, which affects many major cities in Europe. Tourism boards have an opportunity to build the infrastructure of smaller destinations around hostels. For instance, at Hossegor, a surf destination on the Atlantic that has never been considered a youth tourism hotspot, Accor Hotels built the first Jo&Joe hostel, which is helping to put it on the map for backpackers.

There are numerous small destinations near larger tourist areas where visitors can be diverted from crowded hotspots to lesser-known locales. Evidence shows that if a hostel is well-designed and offers good facilities, it can attract a new type of visitor. As a digital nomad, I see this growing segment as vital; if there's a clean coworking space and reliable Wi-Fi, people are more likely to spend extended time in these locations.

In particular, many travellers are seeking experiences in nature, moving away from the hustle and bustle of big cities. There's a tremendous opportunity for tourism boards to engage with and attract these travellers.

For example, Portugal faces a challenge with overcrowded cities like Lisbon and Porto. In response, smaller villages in the Algarve are now creating spaces for young travellers and digital nomads to come and contribute to the local economy.

Tourism boards need to identify the needs of today's youth travellers, as these align closely with the needs of digital nomads. Both groups seek clean, comfortable accommodation with access to nature and adventure sports. This presents an opportunity for hostels to cater to a niche market focused on adventure, as I predicted five years ago.

You're starting to see more adventure-focused hostels, such as the Midgard Base Camp in Iceland, which is in a very remote location. They are the only hostel in that area and offer various adventure sports and tours. This type of accommodation is tailored for adventure travellers who want to explore the surrounding region.

Ultimately, the value lies in these niches.

Stephen: I find this advice invaluable for brands or destinations considering how to strategize for youth travel. It raises the question: how do we plan this out? What do we need and where do we start?

The term 'placemaking' has gained traction over the last 5 to 10 years. What you're suggesting is that youth travel shouldn't happen by accident; instead, destinations need to deliberately plan for it. If we're going to have a hostel, what additional infrastructure is necessary? For instance, there's no point in having a hostel in Western Iceland without nearby adventure sports or transport options. Travellers also need access to places to eat and socialize.

Ultimately, it's essential to consider how to make the destination accessible and appealing.

Kash: I believe Europe is doing an exceptional job of promoting youth travel, particularly through initiatives like the Interrail pass giveaway, which I think was brilliant. If I recall correctly, there was a giveaway aimed at individuals aged 16–19, offering free internal passes for young travellers pre-university.

I had a couple of journalist friends who knew about my book on hostels and they purchased it for their kids, who were planning a big adventure around Europe. One friend reached out to ask for tips on where her son could go and what activities I would recommend.

The Interrail pass is a prime example of how providing opportunities for youth travellers can enhance their experiences. Imagine if we had access to something like that during our time; it would have opened up a world of possibilities.

Stephen: Interrailing was an incredible experience, offering a remarkable sense of freedom in travel. I was fortunate enough to travel at a younger age and that experience shaped my perspective on exploring new places. Considering the sustainability of travel, trains are a much more eco-friendly option. This could potentially become a future trend where more travellers choose trains over planes.

Kash: In Germany, we have an excellent initiative called the Deutschland Ticket, which allows users to pay a flat fee of €49 per month for access to almost all types of trains, excluding the high-speed ones. This has opened up a world of travel opportunities for people of all age groups.

If governments were to invest in creating a youth mobility pass, allowing young people to travel for a low fee or even for free, it could be a fantastic way to attract youth travellers to various destinations. Implementing smart ideas like this, especially during the low season or winter months, would not only promote tourism but also optimize resources without overwhelming local infrastructure.

Stephen: Being smart and agile is an invaluable strategy for destinations! As we explore the long-term benefits of youth travel in this book, I'm interested to hear if you've seen any destinations that excel at encouraging return visits and fostering a welcoming environment for travellers.

For instance, I've been fortunate enough to visit Cambodia three times. By my third trip, the locals recognized me and were genuinely surprised that I kept coming back. I explained that there's truly no other place on earth where you can witness such a breathtaking sunrise as at Angkor Wat. That unique experience was a key factor in my decision to return; and the ease of obtaining my visa – conducted entirely online without the need to visit an embassy – made the process even smoother. It's the combination of unforgettable experiences and accessible logistics that keeps travellers like me coming back. I'm curious if you've had similar experiences or insights in this area.

Kash: A prime example of an inviting destination is Thailand, which facilitates travel by offering a 30-day visa on arrival, soon to be extended to 60 days. This strategic move encourages travellers to stay longer, reducing the urge to hop to other countries. Streamlining the visa process minimizes worries for young travellers, making their experience smoother and more enjoyable.

The consistency of experiences, like the one you described at Angkor Wat, is vital. Despite changes over the years, the magic of witnessing the sunrise remains largely intact, allowing visitors to enjoy the beauty without overwhelming crowds – provided they arrive early. Asia excels in creating memorable travel experiences, even amid the challenges of over-tourism. Destinations here are often so enchanting that travellers feel compelled to share their experiences with friends and family long after their return.

Recently, I visited Mexico, which had long been on my list. It struck me as the Southeast Asia of the Americas, with friendly locals and a welcoming atmosphere. Despite language barriers, the warmth and accessibility of the culture make it a favoured destination for travellers of all ages.

Portugal is another standout, known for its innate hospitality towards visitors. During my time living in Madeira, I witnessed how deeply ingrained this welcoming spirit is among the locals. Tourism is the primary source of income on the island and that influences their natural friendliness.

Creating a culture of hospitality is essential for destinations looking to attract visitors. It requires a commitment to making travellers feel at home, giving them a sense of belonging. While there's no magic formula, focusing on robust infrastructure – like a thriving hostel scene – can significantly enhance the travel experience. Mexico, in particular, is experiencing rapid growth in backpacker tourism, making it a top choice for youth travellers due to its affordability and vibrant culture.

Planning for youth travel success

Stephen: To achieve long-term customer loyalty and maximize the lifetime value of youth travellers, destinations must prioritize creating exceptional experiences tailored to this demographic. For instance, Madeira, traditionally known for attracting older travellers, has successfully evolved into a more youthful destination. This transformation has been facilitated by strategic adjustments in marketing and infrastructure, making the island more accessible and appealing to younger audiences.

Kash: When I lived in Madeira about seven or eight years ago, people would often say, 'Only your parents go to Madeira – young people don't visit.' However, I always believed in its potential. In the last few years, Madeira has significantly transformed its marketing strategy. They've made a concerted effort to attract journalists and content creators, showcasing the island's unique experiences.

What's impressive is how they welcomed influencers and creators, enabling them to capture those 'wow' moments. In today's world there's an insatiable appetite for fresh experiences, and the media is equally eager for inspiring content. Once a destination gains traction with a few viral videos, it can quickly become the next big thing.

Unlike 20 years ago, when Lonely Planet was the definitive guidebook that shaped travel fortunes, today's travellers frequently turn to social media – particularly TikTok – to research potential destinations. The way youth travellers plan their trips is evolving rapidly. They seek authenticity and memorable experiences, and platforms like TikTok offer a more engaging glimpse into what a place has to offer before they decide to visit.

Stephen: That's a fantastic perspective to conclude on. It's crucial for those involved in making decisions about destinations and marketing strategies to understand the evolving landscape of social media. Many may not even be aware of what TikTok is, yet it's becoming a pivotal platform for reaching younger travellers. By leveraging TikTok and similar platforms, destinations and brands can significantly enhance their appeal and engagement with youth travellers. It's essential to adapt to these changes to attract and retain a younger audience.

Kash: Many destinations are still fixated on traditional media like legacy publications, newspapers and *National Geographic*. However, today's youth travellers don't engage with those formats; they turn to platforms like TikTok and Instagram for inspiration. Unfortunately, numerous destinations have yet to recognize this shift and fail to invest in these vital channels.

Conclusion

This conversation with Kash highlights the importance of understanding and meeting the evolving needs of youth travellers. From creating unforgettable experiences to utilizing modern marketing strategies, both travel brands and destinations must adapt for long-term success. Hostels have emerged as a crucial element of the youth travel experience and their quality can significantly attract younger travellers to a destination.

Once overlooked, hostels now deserve recognition and support from governments, local councils and the broader tourism sector as an integral part of the tourism infrastructure. Embracing this shift in perspective will enhance the overall appeal of destinations to the youth market.

KEY TAKEAWAYS

1 **Evolution of hostels:** Hostels have shifted from being cheap, basic accommodation to places that cater to the lifestyle needs of young travellers. Today's hostels often include amenities like coworking spaces, bars, gyms and rooftop yoga, attracting remote workers and digital nomads. Hostels are now viewed as more than just places to sleep; they are community hubs where travellers can connect with others and engage in enriching experiences.

2 **Youth traveller preferences:** Youth travellers are increasingly seeking authentic cultural experiences rather than staying in isolated tourist environments. They want to engage with locals and immerse themselves in the culture. Destinations like Thailand and Mexico are highly favoured due to their affordability, safety and accessibility. These places offer adventure activities and vibrant hostel scenes that appeal to young travellers.

3 **Destinations' role in supporting youth travel:** Destinations that invest in youth travel infrastructure, such as hostels, are likely to attract more young travellers. Smaller towns and cities can benefit from this by offering unique experiences that divert travellers from overcrowded tourist hotspots. They should recognize the long-term value of youth travellers, who spend longer in the area and contribute more directly to the local economy compared to luxury travellers. Destinations like Australia and Germany have been successful in implementing youth travel campaigns.

4 **Future trends in youth travel:** Youth travel is evolving, with a focus on sustainability, adventure and unique experiences. Destinations that cater to these trends will have an advantage in attracting young travellers. Digital nomads are becoming a growing segment of youth travellers. Hostels that provide good coworking spaces and internet access are likely to appeal to this group. Infrastructure development in smaller destinations around hostels can help address over-tourism in major cities.

5 **Importance of digital marketing and social media:** The rise of platforms like TikTok and Instagram has transformed how young people plan their travels. Social media has become a key source of travel inspiration, so destinations must invest in these channels to attract youth travellers. Traditional forms of marketing, such as guidebooks and print media, are becoming less relevant to younger audiences. Destinations that fail to engage with digital marketing are missing out on valuable opportunities.

5

Investing in youth travellers as key drivers of economic growth

In this conversation, Sally Cope, the former head of Tourism Australia for the UK and Europe, recounts her own journey from youthful exploration to a seasoned career in the tourism sector, offering a first-hand perspective on how youth travel experiences can shape lifelong connections. She sheds light on Australia's robust approach to youth engagement, from the enduring appeal of the Working Holiday visa to innovative marketing campaigns designed to resonate with young travellers across the globe.

The interview unpacks the challenges and successes of positioning Australia as a long-haul yet must-visit destination for young adventurers. Sally emphasizes the importance of strategic storytelling, tapping into nostalgia and harnessing modern marketing methods to keep Australia relevant and appealing in a competitive global market. Her reflections provide valuable lessons for destinations looking to foster not just visits but lasting bonds with travellers who could evolve into future advocates, repeat visitors or even contributors to the destination's community and economy.

Through this interview, you will gain a deeper understanding of the intricate relationship between marketing strategy, cultural affinity and economic investment that underpins successful youth travel initiatives. More importantly, this chapter highlights the critical value of seeing beyond the immediate gains of tourism and focusing on nurturing relationships that yield long-term dividends.

Stephen: So, what are you up to these days?

Sally: I never did the big backpacking travel journey as an 18-year-old. So, I've decided to do it in my mid 50s instead. I'm having six months off and I've just walked across the UK doing the Coast-to-Coast Trek, and next week I'm off to Europe for a couple of months, so I'm looking forward to that.

Stephen: As someone who backpacked your homeland of Australia and is now at 43 hoping he would be able to do it again, you give me hope that I will. Maybe one day!

Sally: It's never too late, I've decided.

The journey from Working Holiday visa to economic driver

Stephen: In my discussions with other people for this book, Australia has come up a lot, in a positive way. What would be great to hear is your view and your knowledge of how important youth travel is to Australia as a destination.

Sally: The Working Holiday visa programme has been an integral part of Australia's travel landscape since the 1970s. It's been hugely successful in building long-term, people-to-people connections between Australia and other countries worldwide. The programme has not only allowed international visitors to come and spend extended periods in Australia, but it's also given Australians the opportunity to travel abroad for similar experiences.

Geography has played a big role in this. Australia is quite distant from most countries, especially those in the Northern Hemisphere. It's a long-haul destination that requires time to explore properly, which is why it appeals to young travellers who are carving out time for a significant trip. For many, travelling to Australia is seen as a challenge – it's as far away as you can go, geographically speaking, and it feels very different due to its vast and varied landscapes.

However, particularly for British travellers, there's also a sense of familiarity because of the historical ties between the two countries. There's a shared mindset and cultural connection, making it feel less intimidating than other distant destinations. So, while it's a bold move to travel so far, it's not as daunting as it might seem.

TIP

Tapping into links via heritage is not just for the older demographic. Familiarity combined with popular culture can attract young people to your destination.

Media and popular culture: Shaping perceptions

Stephen: I certainly felt a sense of familiarity during my visit at age 16, largely due to my exposure to Australian culture through the television show *Neighbours* and the film *Crocodile Dundee*. Upon returning for my Working Holiday visa, I found myself even more immersed in Aussie culture, especially with the 2000 Olympics, which made me feel safe and comfortable as I planned my gap year.

Sally: It's important not to underestimate the influence of media, which continues to shape perceptions today. While *Neighbours* may not hold the same prominence, many people are inspired to undertake similar journeys by following individuals on social media who are currently travelling around Australia. This connection with real experiences can be quite motivating.

Youth travel is also vital for the country, as it represents a significant sector within the tourism industry. Stakeholders within Australian tourism recognize the immense value of working holidaymakers, who tend to stay for extended periods, contributing to the economy as high-yield travellers. They spend considerably and help fill essential seasonal roles in various industries, including agriculture and tourism, which is crucial for the vitality of regional Australia.

Youth travel: A key component of Australia's economy

Stephen: It seems that there is a comprehensive focus and budget allocation for youth travel from a top-down perspective. Do you believe this emphasis on youth travel is deeply ingrained in Australia's tourism strategy?

Sally: Indeed, throughout my 30 years in Australia's tourism industry, this focus has been evident across various sectors. From a product standpoint, the value of the workforce is recognized, particularly in remote resorts that recruit working holidaymakers annually. Additionally, there is a substantial youth tourism sector that offers arrival packages and support services, helping young visitors integrate into a community of like-minded individuals. This fosters connections and provides essential travel advice, greatly benefitting hostels and adventure tourism operators nationwide. Furthermore, transportation options – whether by bus or plane – are accessible to enable young travellers to explore the vastness of Australia.

Stephen: My experiences included a mix of Greyhound coaches and organized tours, such as the Great Ocean Road and a guided trip to Uluru, which were highlights of my time in Australia.

Sally: I worked at the Ayers Rock Resort in Central Australia for several years and the transient workforce there greatly benefits from the regular influx of working holidaymakers. Uluru, or Ayers Rock, is located right in the middle of Australia and is very isolated, so these young people are essential to the success of resorts like where I worked.

The long-term value of youth travellers

Stephen: One of the themes of my book is the long-term value these young travellers bring. Is this something Australia tracks or understands when investing in youth travel, and how is the success of such investments measured?

Sally: There is a fundamental understanding of this value. In the mid-2000s, I participated in a global roadshow with Australian industry colleagues to rebuild confidence in travel following the September 11 attacks. During this time, the youth market was among the first to recover, demonstrating resilience. I recall our Federal Minister for Tourism highlighting the youth market as a powerful investment in the future. These travellers often form enduring connections with Australians, which can lead to long-term business partnerships or friendships that transcend borders.

I exemplify this theory; I came to the UK on a Working Holiday visa in the 1990s and have maintained strong ties, feeling a sense of belonging in both countries.

Stephen: It is refreshing to hear such a deep understanding from a minister regarding this sector, as that is not always the case. From the inception of the Working Holiday visa, has Australia tracked the inward investment and its impact on the economy?

Sally: Yes, there has been a clear distinction between Working Holiday visas and skilled labour visas. Many may misinterpret a Working Holiday visa as a stepping stone to immigration, but its true intention is to provide an extended opportunity for cultural immersion. The ability to work while travelling allows individuals to stay longer and continue exploring.

Success is measured by the number of visitors, the duration of their stay, the rate of extensions for second years and their spending patterns.

For Tourism Australia, the most critical metric is the economic contribution of these young travellers, which is often more substantial than that of luxury travellers who may spend a lot in a short time.

Stephen: Similar to my experiences on Greyhound buses in remote areas, young travellers today still embody a sense of adventure, even with laptops replacing traditional travel guides. Many destinations lack luxury offerings, indicating that young visitors are likely to spend their money in ways that luxury travellers might not. Additionally, with their increased awareness of environmental impact, they may travel more sustainably than those on brief luxury trips.

Sally: Youth travellers prioritize experiences over material goods. While they may not spend as much on nightly accommodation, they invest significantly in unique experiences, such as skydiving or dining under the stars at Uluru. This focus on memorable, once-in-a-lifetime experiences aligns perfectly with the values of today's youth travellers.

Lessons from a long-haul destination

Stephen: As I noted earlier in the book, we are now in an age that favours 'experiences over possessions'. Ownership of material goods, such as a home, is less critical than the experience of sleeping under the stars at Uluru. I identify as a 'geriatric Millennial', born in December 1980 and straddling the line between generations. My travels included using an emergency phone and frequenting Internet cafés, positioning me at the cusp of the digital age. Nevertheless, experiences like cooking and camping under the stars at Uluru remain profoundly important to me.

Regarding marketing a long-haul destination, what challenges arise in promoting such a distance to youth travellers and how have these been addressed?

Sally: At Tourism Australia, we conducted extensive research to identify barriers preventing people from considering travel to such a far-off destination. Over the years, perceptions have shifted and I believe they will continue to evolve. Eight to ten years ago, pre-Covid, there was a pervasive fear associated with gap years, as individuals worried about falling behind in their careers. This perception hindered travel decisions.

In a post-Covid world, where travel opportunities were restricted, there is now a sense of urgency to seize the day. However, the barrier of distance remains, with concerns about missing family and friends. Our challenge as a tourism board has been to reframe these thoughts positively.

Previously, when the focus was on career advancement and potential setbacks, our counter-argument highlighted that travel can enhance one's career through valuable life experiences. Returning travellers often find that their time abroad enriches their resumes.

Similarly, in the current climate, when young people express concern about being far from home, we encourage them to embrace the notion of prioritizing their own experiences and personal growth.

As marketers, our task is to identify barriers and determine whether they are perceived negatively or can be transformed into a positive narrative.

Stephen: Ultimately, it's a challenging question because Australia cannot be made geographically closer to mainland Europe. During Covid, I had an Australian colleague who experienced the contrast between Queensland, where travel was thriving, and Melbourne, which faced frequent lockdowns. Fortunately, Australia eventually reopened to international travellers, quickly recovering from initial concerns.

Sally: Interestingly, the youth sector was the first to rebound post-Covid, mirroring trends observed after 9/11. Australia is currently experiencing an unprecedented influx of working holidaymakers, reaching peak levels, which is remarkable. Despite rising airfares, young people are still choosing to travel. For many, this is their first significant trip and they tend to focus on the experience rather than comparing airfare costs to those from 5 or 10 years ago. They are simply embracing the opportunity to go.

Stephen: In recent years, several adjustments have been made to various working holiday schemes, including the one between the UK and Australia, which has raised the upper age limit from 30 to 35. This change potentially opens the door for an additional 5 million young people aged 30–35 to travel to Australia, creating a significant new market. I imagine this has had a highly positive impact on Australia and it seems the country has embraced this opportunity.

Sally: This change follows a similar increase already implemented in Ireland and reflects a gradual expansion for all countries with reciprocal visa agreements. The response in the UK was quite interesting; it drew significant attention and opened opportunities for millions more British citizens. Historically, we have found that those who undertake working holidays in Australia tend to be recent university graduates, typically in the transition between completing their degree and starting their first job. However, post-Covid, we may be witnessing a shift towards mid-career sabbaticals, with individuals considering taking 6 to 12 months off or travelling between jobs.

This shift presents an exciting challenge from a marketing perspective. We must consider how to reach these individuals effectively. For example, should we engage with them on LinkedIn? Someone even suggested targeting dating websites, playing on the idea that someone might just have gone through a breakup and is ready for a fresh start. It raises the question: what life events serve as catalysts for the decision to travel?

Stephen: It's encouraging to hear your response. When the announcement was made last June, Australia swiftly began marketing and introducing new concepts. Unfortunately, the British government did not implement similar efforts to attract Australians to the UK, which is quite disappointing.

The British youth travel industry, including hostels and adventure travel providers, sees this as a missed opportunity to significantly boost the sector. Ideally, we would hope for a reciprocal approach that fosters mutual travel and exchange in equal numbers. While it's understandable that the larger population in Britain means exact parity may not be feasible, the lack of marketing efforts presents a challenge. Awareness is crucial, and without it we face obstacles in encouraging travel in both directions.

For the UK youth sector, it seems that not as many Australians are aware of the age increase to 35, compared to how quickly Brits learnt they could take that mid-career break.

Sally: I'm flipping the question back to you now: do you think the Australian youth traveller is taken for granted? In Australia and New Zealand, it seems almost like a rite of passage for young people to spend a year travelling abroad, and this tradition persists to this day. I'm curious whether there's an assumption that they will continue to do so, perhaps leading to the perception that their participation is guaranteed.

Stephen: I believe we do take youth travellers for granted in the UK. This tendency often occurs in the travel industry when brands and destinations with a rich heritage become complacent, expecting visitors to simply show up. This is evident even in the luxury sector, where there's a belief that US tourists will always come, which I find naive. There are now many other easily accessible destinations offering direct flights from the US and, over time, visitor volumes could decline.

I've also noticed through friends that more Australians are choosing to go to the US for their working holidays, as it has become more accessible. This trend represents a loss for the UK, especially since Australians in New York are only about six hours away, making it feasible for them to visit the UK while based in the US.

So, yes, I do think youth travellers are taken for granted in the UK, but this is more of a 'Brand Britain' issue rather than solely a youth travel concern. Additionally, I believe the travel industry lacks comprehensive data to fully illustrate the power, volume and value of youth travel – not just in the UK but in other destinations as well. This absence of data hinders tourism boards and governments from recognizing the economic advantages of investing in this sector.

In contrast, Australia tracks data effectively, as do several other destinations. This capability aids in developing marketing strategies that combine creativity with data, resulting in a significantly greater return on investment.

Sally: I must admit, I used to worry about the significant investment we made in marketing to this audience. We always had to be clear and intentional in ensuring our messaging was distinctly Australian, rather than simply promoting the sector. You're right – many destinations don't market with the same vigour as Australia, so it was essential for us to communicate what makes Australia unique. Our campaigns had to convey that these experiences could not be found anywhere else, which was vital for achieving our objectives.

Stephen: I've done some lecturing, using examples of destination marketing to illustrate how straightforward it can be. A classic example is the cultural impact of *Crocodile Dundee* and the phrase 'Shrimp on the barbie', which significantly raised awareness of Australia as a destination. One advertisement I reference is from around 2004, featuring Delta Goodrem singing 'I Can See a Rainbow' alongside a kaleidoscope of images showcasing Australia. It lasts no more than a minute but is incredibly powerful – watching it can give you goosebumps. If an advertisement evokes that kind of emotion, you know you're on the right track!

Australia is fortunate to possess so many iconic views and natural wonders, and I believe other destinations could learn a lot from your approach to emotional marketing – whether it elicits goosebumps or laughter. This emotional connection is crucial.

Another example I use is from the Hong Kong Tourist Board, launched in 2020: 'Let's Rediscover Hong Kong – Our Home.' Given the context of Covid lockdowns and the geopolitical challenges at that time, it was a compelling campaign. It highlighted the people and key sites, complemented by music from a local pop star, adding a unique flavour. In just one minute, it effectively showcased the destination and conveyed why people would want to visit.

Emotional marketing, whether through television or platforms like YouTube, truly attracts young travellers – and even older ones like me! However, it also requires significant creativity from marketing teams to develop such impactful campaigns, as well as support from the wider industry and community to provide the powerful content needed to make it all sync together.

Sally: I believe social media has made things more challenging as well. When engaging with a younger audience, there seems to be a greater emphasis on endorsements from friends or admired figures rather than from brands themselves. This shift makes it difficult to present authentic messaging or to cut through with polished brand advertisements. For destination marketing aimed at young travellers, there's significantly more value in collaborating with quality content creators on social media who are authentic.

From our perspective at Tourism Australia, it was essential to connect with real British individuals currently living in Australia who were genuinely enjoying their experiences and sharing their lives online. Tapping into that energy and sharing their content felt much more authentic and resonated more strongly with the audience.

TIP

Word of mouth for young travellers now happens through social media, so it is essential you are accessible to this sort of media, otherwise you will be ' lost in the noise'.

Stephen: Was there ever consideration at Tourism Australia about the long-term journey of working holiday travellers? Specifically, once they return home, what marketing strategies were in place to maintain that connection?

Sally: Yes, absolutely. Those individuals who return home are invaluable advocates. If we can harness that potential, it becomes extremely valuable, especially now. Someone who has had a great experience abroad can serve as an evangelist, sharing their journey and insights.

We explored various ways to connect with these returnees and collaborate with them to share their content. When someone else tells your story, it carries far more weight than any branded message.

Creating lifelong ambassadors

Stephen: Experiencing a different culture – something that smells, looks and feels unique – creates a profound emotional connection. It's like visiting a new restaurant for the first time; when you encounter a new cuisine or a fresh take on a familiar dish, it can leave a lasting impression. You truly bond with your friends over an exceptional meal or an unforgettable holiday, wanting to share those experiences.

Do you find that it may be challenging to track? For instance, those backpackers from my era in the early 2000s, who contributed to Australia's recovery after 9/11 – does Australia have any insight into whether they returned? Have they opted for five-star hotels, played golf, or even brought their children along, encouraging them to embark on similar journeys?

Sally: I'm not aware of any data collection that directly establishes that connection, so no, it isn't specifically measured. I would assume the number is high, but we don't have concrete figures.

However, the repeat visitation rate from the UK to Australia is indeed very high, exceeding 50 per cent annually. Many people come and often they realize they missed out on destinations like Tasmania or Kimberley. This prompts them to return and create entirely new travel plans.

TIP

People who have spent a year or two on a Working Holiday visa will often come back and explore the same destination further. It is therefore important to make sure you re-market to previous travellers to your destination and make sure you keep in touch with them!

Stephen: Well, a 50 per cent return rate is quite impressive. For those who may read or listen to this and are not in the tourism industry, that figure signifies strong interest and engagement. It reflects a successful connection with travellers who want to return.

Sally: Yes, absolutely. That distinction between the UK and Australia is quite notable. It all comes back to the strong cultural, historical and sporting ties between the two countries. There's no denying that the longevity and success of the reciprocal Working Holiday visa programme have significantly contributed to this connection.

Stephen: It's fascinating to consider how destinations could track this in the future. For instance, when people apply for a visa, they could indicate whether they previously participated in a working holiday programme.

That leads me to my next question. Given your experience with various campaigns for Tourism Australia, is there one that stands out as your favourite, and why?

Sally: One campaign that stands out for me, particularly in terms of clever and witty advertising, was created around 2018–19, prior to Covid. It addressed the decline in the sector due to concerns about falling behind in careers. We developed a brand called 'Australia Inc.' as if it were a company, and produced a tongue-in-cheek recruitment video for working holidaymakers. The commercial highlighted the unique and diverse jobs available across the country and was executed very slickly.

The rollout included a bit of ambush marketing. We set up 'Australia Inc.' stands at recruitment fairs, targeting university graduates, who we knew were most likely to consider taking a gap year. While they were meeting with major accounting firms and other potential employers, they encountered our tourism stand, which was designed to resemble a business.

This campaign resonated well and effectively addressed the concern that young people had about falling behind. I was quite proud of it, as it performed very successfully.

However, I don't think that approach would work today; the mindset and barriers have shifted significantly. Currently, I believe Tourism Australia is on the right track by leveraging authentic content created by real working holidaymakers on the ground. This approach feels less manufactured, yet it is likely just as, if not more, effective because it is relevant to today's audience.

> **TIP**
>
> Storytelling is so important, and making sure that whatever you are marketing has a 'hook' and a story means that you will have greater success compared to a dull and static campaign.

Stephen: It's great to hear that. Tracking the success of any marketing campaign is essential. You need to know who your target audience is and ensure that your messaging is relevant to them, regardless of the medium. As all marketers understand, there's always pressure from management to demonstrate the return on investment (ROI).

Some of that return is tangible, such as the number of people signing up for Working Holiday visas. However, there's also the long-term impact – people who see your campaign and think, 'Wow, I'm going to do this when I'm 30 or 32.' While you can't track every influence, the overall effectiveness of the campaign certainly becomes evident over time.

Sally: As a tourism board, it's challenging to track our impact because we're not directly selling anything. However, we can count the number of visas issued over time, and we utilize the International Visitor Survey from Tourism Research Australia to track average spending.

Since a working holiday in Australia is often a long-term consideration, our campaigns may spark interest, but it could take up to 12 months before someone actually travels. Then, it may be another year before we truly understand their spending patterns in the country.

In destination marketing of this nature, consistency is key. It's not about achieving a quick sales spike and declaring success; rather, we need to persevere and follow trend lines to ultimately see the results of building that brand.

Stephen: Having worked in marketing, I've always emphasized the importance of considering both the long-term impact of your campaign and the short-term data. While short-term results may not always meet expectations in terms of ROI, if a campaign has sparked the dreaming stage for potential travellers to your destination, that's incredibly powerful.

Tracking this impact can be more challenging, but, as you mentioned, the mechanism with the Working Holiday visa allows you to do that effectively.

Conclusion

This conversation highlights the critical importance of youth travel as both an economic driver and a long-term investment for destinations, particularly highlighted by Australia's comprehensive approach. Youth travellers, from backpackers to working holiday participants, contribute significantly to local economies through extended stays, active participation in experiences and cultural exchanges. This demographic not only brings immediate economic benefits but they also develop enduring connections that lead to repeat visits, referrals and even cross-generational travel. The success of such strategies lies in authentic marketing and understanding the evolving preferences of younger audiences. Ultimately, cultivating youth travel isn't just about short-term gains – it's about creating lifelong ambassadors for a destination.

KEY TAKEAWAYS

1 **Working Holiday visa:** This programme has been integral to building strong, long-term connections between Australia and other countries since the 1970s. It allows young travellers to spend extended periods in Australia, while also enabling Australians to have similar experiences abroad. Australia's distance from most parts of the world makes it a destination that travellers, particularly youth, spend significant time exploring.

2 **Familiarity and adventure:** For British youth, there is a sense of cultural familiarity in Australia due to historical ties, but it remains a destination that offers vast, diverse landscapes and a sense of adventure. Media, both past (e.g. *Neighbours*) and present (social media influencers), plays a crucial role in shaping perceptions of Australia as a desirable destination for youth.

3 **Youth travel's economic importance:** The youth sector is vital to Australia's tourism industry. Working holidaymakers tend to stay longer, spend more and contribute to the economy by filling seasonal jobs in industries like agriculture and tourism. The long-term value of these travellers is significant, as many maintain connections with Australia, which can turn into business relationships or repeat visits.

4 **Challenges and shifting perceptions:** A challenge for marketing long-haul destinations like Australia has been overcoming fears of distance and isolation. However, post-Covid, young travellers seem more willing to seize the moment and prioritize travel experiences. Australia has worked to shift perceptions by emphasizing the personal growth and career benefits of travel.

5 **Marketing campaigns:** Sally highlights some notable campaigns, such as 'Australia Inc.', which cleverly addressed concerns young people had about taking a gap year. The campaign focused on promoting the diversity of job opportunities available across Australia. More recently, authentic content created by real travellers and shared on social media has become a key strategy in connecting with the youth market.

6 **Long-term impact:** Tourism Australia tracks the economic contribution of young travellers through visas and spending data. However, the emotional and cultural impact – such as the lifelong connections built during these travels – is harder to quantify but equally important. Sally emphasizes the importance of creating authentic, experience-driven marketing to resonate with younger travellers.

6

Authenticity, safety and growth

Capturing the youth travel market

In this interview with Nick Pound, co-founder of World Nomads, a company renowned for its innovative approach to travel insurance, we delve into the critical role youth travel plays in shaping brands and destinations. Nick highlights the importance of building trust, staying relevant in a highly competitive market and creating meaningful connections that resonate with younger audiences. This chapter explores the interplay between experience, safety and authenticity in youth travel, and how these elements can transform a transactional product into a life-enhancing brand.

Stephen: Nick, could you start by explaining what World Nomads does as an organization?

Nick: Absolutely. World Nomads has been around for just under 23 years. We were established in Australia with a simple yet ambitious vision: to create a travel insurance product that could serve people travelling from anywhere to anywhere. To use an analogy, think of us like a can of Coke – it tastes the same whether you're in the US, Canada, the UK, Australia or South Africa. We wanted our travel insurance to have that same universal accessibility and consistency.

When we started, the marketplace was dominated by travel insurance distributed through traditional travel agencies. These products were comprehensive but catered to more conventional travellers. We saw an opportunity to address the needs of a different audience – backpackers, adventure travellers, independent explorers and young people. That was our foundation, and while we've evolved as a brand over the past 20 years, those roots remain central to our identity.

At our core, we're a travel insurance business. But we quickly realized that insurance, on its own, isn't the most exciting product – it's essentially about risk management and, let's be honest, it can feel pretty dry when viewed in isolation. So, we decided to build a brand around it, one that resonated deeply with our target audience and the aspects of travel that matter most to them

---

TIP

You don't always need to have the most exciting product in the world to break into the youth sector. Authenticity and a brand ethos that means something can elevate you from the crowd and connect with your core demographic – what is your USP aimed at youth?

Stephen: So, creating a strong brand was key to standing out. How did that shape the direction of World Nomads?

Trust and safety as competitive advantages

Nick: The brand allowed us to go beyond the insurance product itself and add key value for our customers. We started incorporating elements that spoke to the adventurous spirit of our travellers.

For example, we introduced philanthropic travel initiatives, providing opportunities for travellers to give back to the communities they visited. We launched scholarships – supporting budding travel writers, filmmakers and photographers to hone their craft and gain valuable experience. Safety content became another pillar, equipping travellers with vital information to navigate their journeys with confidence.

Back when blogging was a brand-new phenomenon, we embraced it. We created content that engaged and inspired our audience, sharing stories that went far beyond the transactional nature of insurance. We even developed job services to help young travellers sustain their adventures. In essence, we built an entire ecosystem of brand attributes that allowed us to connect with travellers on a much deeper level. We set out to share this wealth of knowledge accumulated within the World Nomads community, to inspire and educate our customers.

Authentic experiences drive success

Stephen: So, World Nomads became much more than just an insurance provider. It sounds like you created a lifestyle brand that aligned with the values and aspirations of your audience.

Nick: Exactly. We didn't want to be just another provider of 'risk paper', as we sometimes call it. We wanted to be a trusted companion on the traveller's journey. By positioning ourselves as more than an insurance company, we were able to stand out in the crowded travel space and build lasting relationships with our customers. It also has the added benefit of providing another avenue with which potential customers might find our brand and product, whether they're in the early stages of planning a trip and are looking for inspiration, right up to the point of departure when they might be doing some last-minute research on their chosen destination.

Stephen: That's really interesting, Nick. One of the recurring themes in this book is about creating an experience in every part of travel. Travel insurance, while essential – whether someone is studying abroad for a semester or backpacking for a year – has traditionally been viewed as a compliance-driven product. How do you make something like travel insurance part of the overall travel experience, rather than just another piece of paperwork?

Nick: It's a great point, Stephen. Back in the day, an Australian politician called Ian Kemish famously said that if you are unable to afford travel insurance, then you are unable to afford to travel.[1] It's a sentiment that has long resonated in the Australian market. Australians are among the most insured travellers globally, largely because of the nature of their geography. Australia is far from most other places, with the closest international destinations being New Zealand, Bali and Southeast Asia – and even those involve long-haul flights. Because of this, Australians are well-educated about the necessity of travel insurance. It's ingrained in their culture; the message is loud and clear: *Don't leave home without it.*

But when it comes to the wider youth market, behaviours around travel insurance vary significantly by country. For instance, in the UK, travel insurance wasn't something young people typically considered. If you were hopping across to Europe, you had your reciprocal health care card and as a 20-something you felt indestructible. Sitting by a pool for a week didn't seem to carry much risk, so why bother with insurance? It simply wasn't on the radar.

It's taken decades for travel insurance to be recognized as a core part of youth travel planning. Many of the big youth travel brands, which were founded 30 to 40 years ago, have contributed to shifting that mindset. More recently, events like Covid-19 have pushed insurance from an afterthought to a top priority on trip-planning checklists. Before the pandemic, travel insurance might have ranked seventh or eighth in importance; now, it's arguably in the top three, alongside tickets and accommodation.

A survey commissioned by World Nomads in 2024 revealed some interesting insights into the mindset of UK travellers. More than a quarter of Brits wouldn't know what to do if they got sick abroad. As many as 60 per cent are worried about falling sick overseas. And over a third of Brits have experienced a medical situation while travelling. Nearly half (48 per cent) of Brits are worried about the cost of treatment abroad. And 43 per cent of travelling Brits have experienced flight delays or cancellations in the past year.[2]

So, we know how to provide a product that can help alleviate these burdens for travellers. The key is getting the relevant information to our key audience and, crucially, making the value proposition easily understood.

Stephen: That's a huge shift. It seems like the perception of travel insurance has changed dramatically in recent years.

Nick: Absolutely. Covid-19 was a major inflection point. It forced travellers to confront the unexpected. What if the world shuts down again? What if my trip gets cancelled or delayed? Those questions put travel insurance top of mind. The irony is that the product itself hasn't changed much over the years – Covid-specific coverage aside. What has evolved is the global audience's understanding of its importance.

Take the US market, for example. Traditionally, Americans mostly travelled to nearby destinations like Canada or Mexico. But now, they're exploring the world more than ever before. The US has become one of our largest outbound markets. A decade or so ago, Lonely Planet ran a brilliant campaign encouraging Americans to get passports. At the time, a surprisingly small percentage of Americans had one. Lonely Planet's message was simple: the world is accessible. With the rise of low-cost carriers, international travel became more affordable and Americans began venturing beyond their borders in increasing numbers. Today, we see them as one of the most travel-savvy markets, particularly among the under-30 demographic.

Stephen: It sounds like this shift isn't just about awareness but also about evolving consumer habits.

Nick: Exactly. The youth market has changed significantly. The educational piece around travel insurance has become crucial. Travellers today are more informed – they understand the risks of going uninsured, particularly after events like Covid. But it's also about making the product resonate with them. Travel insurance can't just be a sterile, transactional purchase. It needs to align with their values and their sense of adventure. That's particularly true for Gen Z – they're savvy. They're highly literate in terms of research on brands. And brand loyalty doesn't come easy to them. It needs to be earned. It's not enough to talk-the-talk, you must show them that you also walk-the-walk.

At World Nomads, we've worked hard to position travel insurance as more than just a policy – it's a companion for their journey. It's about providing peace of mind so they can focus on what matters: exploring the world.

QUESTION

Does your travel brand or destination have a clear safety plan that comforts both the traveller and their parents?

Stephen: That's fascinating, Nick. I guess part of your journey has been about educating your audience. Over the past 23 years, you must have seen many moments that triggered significant changes in behaviour. Covid was obviously a massive one, but it's unique in that it wasn't just a disruption – it physically stopped people from travelling altogether. Over the years, how have you seen young people's travel habits evolve, and what are the key trends that have shaped the way they travel with – or without – insurance?

Nick: Covid was unprecedented in terms of its impact on travel. It forced people to stay in their countries, even in their homes, for extended periods – a disruption unlike anything else we'd encountered. That said, it's not the only major event that has influenced traveller behaviour over the years.

Looking back, there have been other pivotal moments. For instance, the Bali bombings in 2002 were a significant event for us as a young

Australian company. What stood out to us was the resilience of the Australian youth market. Just weeks after the bombings, when it was deemed safe to return, young Australians were back in Bali, snapping up the deals and helping to reignite the local tourism economy. It's a pattern we've observed repeatedly: the youth audience is incredibly receptive, adaptable and eager to get back out into the world.

From a broader trend perspective, our business has grown alongside the Millennial generation, which was our core audience when we started. Over time, we have evolved with their habits, particularly as Gen Z has entered the market. Being a digital-first company, we've always been designed for the internet age. Gen Z, with their focus on digital engagement, has further pushed us to refine our offerings and strategies.

Travel insurance itself is a straightforward product – you come to our site, get a quote, input your details and your cover can start. But how do you make a simple, transactional product resonate with a digitally savvy audience bombarded with marketing messages daily? The answer lies in building trust and creating meaningful connections.

One way we have done this is by focusing on the quality of our partnerships. We work with premier-tier insurance providers to ensure our customers receive the best coverage. But beyond that, we have developed initiatives that reflect the values of our audience. A pivotal moment for us was in 2004, after the Indian Ocean tsunami. Like many, we wanted to help. Initially, we sent funds to relief efforts, but it felt impersonal – like just dropping money into a bucket.

That led to the creation of Footprints, our philanthropic arm, in January 2005. The idea was simple but powerful: policyholders could add a small donation – for as little as a pound, a euro – to their travel insurance policy and that money would go directly to a specific project. Importantly, we ensured complete transparency. For example, we'd raise $5,000 or $10,000 for a project and once the work was completed – whether it was building a school in Nepal, sanitizing wells in Kenya, or conducting eye surgeries in India – we reported back to our customers. We showed them the tangible impact of their valuable contribution.

This initiative closed the loop on charitable giving, removing cynicism and building trust. It also aligned perfectly with the values of our audience. They saw us as more than just an insurance company; we were a brand that resonated with their beliefs and sense of purpose.

Stephen: That's a great example of turning something transactional into something meaningful. It's one thing to sell a product; it's another to build a brand that aligns with your audience's values.

Nick: Exactly, and Footprints changed the game for us. It not only deepened our connection with customers but also opened doors for business-to-business (B2B) partnerships and expanded our reach in the business-to-consumer (B2C) market. When people searched for travel insurance online – particularly when Google AdWords came into play – our philanthropic efforts made us stand out. Customers didn't just see a company selling policies; they saw a brand that shared their values and was actively making a difference.

By embedding purpose into our business model, we've been able to differentiate ourselves in a crowded market and create a product that resonates deeply with the modern traveller. It's not just about providing a safety net – it's about being a part of their journey, both as individuals and as global citizens.

Collaborations and ecosystems

Stephen: Yeah, and I guess authenticity in youth travel – whether it's in the experiences themselves or the services that support them – has always been essential. Back in 2005, Facebook hadn't yet launched in the UK and Google was only just establishing itself as the number one search engine. When I travelled in 2003 and earlier, and probably when you did too, the youth travel experience had to feel real and genuine to resonate.

That sense of authenticity seems even more critical now. Young travellers often approach their journeys with a healthy dose of scepticism, whether about the accommodation, the flights or other parts of the process. Your approach with Footprints – creating a transparent, quantifiable way for travellers to give back – must have strengthened your brand. But beyond that, it likely encouraged more young people to consider travel insurance, which ultimately benefits them during their journeys.

It's such a vital thing, because no one wants a young traveller to have a life-changing experience go wrong and find themselves unable to get help. I was aware of the Footprints initiative, but I had no idea how it started. That's a fantastic story.

Nick: Thanks, Stephen. And what's even more interesting is how the programme has grown. A while back, we re-engineered the software to allow other businesses to incorporate Footprints into their purchase paths. Another brand we work with in the Australian market has embedded it into their transactions, and across all partnerships, it has raised over $5 million for charitable programmes worldwide.

Initially, Footprints was inspired by the United Nations Millennium Development Goals. Back then, our projects were more focused on basic needs, like repairing schools. For instance, our first initiative involved rebuilding a school in Nepal. But over the years, as the priorities of our traveller base have shifted, so have the projects we support. Conservation and environmental initiatives are now at the forefront.

One of our recent efforts was focused on sea turtle conservation – a reflection of how much more important sustainability and environmental protection have become for today's travellers compared to the Millennial audience we started with. It's fascinating to see this evolution.

To date, we've completed over 270 projects, from clean water initiatives and eye surgeries to conservation programmes like the sea turtle project. It's a testament to our audience and their desire to give back, to leave a positive impact on the world they're exploring.

What's also been inspiring is how Footprints gave rise to our Responsible Travel Manifesto. This framework has become a cornerstone of who we are as a company, reinforcing our commitment to supporting travellers and the destinations they visit. It's helped us stand out in the market for nearly two decades and continues to shape our identity as more than just a travel insurance provider.

QUESTION

Does your brand have a social impact strategy? If it does, do you share this information? Is there more you could do for your community or those less fortunate further afield?

Stephen: I think that's such an important point, Nick. For brands and destinations targeting the youth market, authenticity is non-negotiable. If you make a claim – whether it's about sustainability, social impact or being eco-friendly – you have to back it up with action. It can't be a greenwashing exercise to score quick wins, because today's young people are diligent. They'll research, investigate and call you out if they find inconsistencies.

Not every traveller may prioritize these things, of course, but many do. When a young person donates a dollar or chooses a brand that claims to be environmentally responsible, they expect to see tangible evidence of impact. It's one of the challenges in travel today – destinations and brands making bold claims without delivering on them. For instance, a hotel might market itself as the 'most eco-friendly in the world', but when you arrive there's single-use plastic everywhere, no recycling programme and no meaningful effort towards sustainability.

It's clear that your decision to create an initiative like Footprints wasn't just about looking good – it was a genuine move and it's something you've stood by over the years. That's rare and commendable in an industry where talk can sometimes outweigh action.

Nick: Absolutely, Stephen. The Gen Z traveller has raised the bar significantly. They don't make impulsive decisions – they're incredibly thorough in their research. Studies show that this generation references around 20 different sources before making a travel decision.

And they don't just rely on traditional platforms – they lean heavily on peer-to-peer reviews, influencers and social proof. The tools at their disposal today are light-years ahead of what we had when we were travelling 20 to 25 years ago. I see it first-hand with my daughters, who are 21 and 19. Their approach to planning trips or making any major purchase is fundamentally different from ours. They've got this library of content and resources at their fingertips, far beyond what we could have imagined back then.

This is why authenticity and trust are paramount. Over the last 20-plus years, we've built a pedigree as a brand that does the right thing, not just for marketing purposes but because it's woven into our identity. This has helped us resonate with a generation that values integrity.

Outside of industry recognition, which is great, the real measure of success for us is the trust our customers show when they repeatedly choose us. They buy our product, knowing they are in safe hands. That trust, earned through years of doing the right thing, is the ultimate testament to our work. It's what motivates us to keep evolving, meeting the needs of a highly discerning and research-driven market.

Stephen: Going back to the theme of peer-to-peer communication, what was once word-of-mouth in pubs or university halls has now shifted online.

Is World Nomads conscious of being part of that digital conversation? And how do online reviews play into your operations? I'd imagine they're incredibly important for a product like travel insurance. People tend to share their bad experiences much more readily than their good ones, but when they do share positive stories, it can be extremely powerful.

The digital evolution of youth travel

Nick: You're right. Peer-to-peer communication has transformed in the digital age, and it's something we've had to evolve alongside. We sell insurance online, so from the moment a customer buys a product to when they're travelling their interaction with us often continues in that online space. But the reality of our industry, as you pointed out, is that we typically only hear from customers when something goes wrong.

In the world of travel insurance, those 'bad experiences' usually translate into claims. Broadly speaking, claims fall into two categories: medical emergencies and inconvenience claims. The latter includes cancellations, lost luggage or minor medical issues. Medical emergencies, however, are where the stakes are highest. These situations require immediate, organized and compassionate responses.

To handle these emergencies effectively, we've partnered with trusted organizations who specialize in providing emergency assistance at any time of day, no matter where you are in the world. From New Zealand to the UK and stretching all the way to the west coast of the US, we have partnered with centres that ensure our customers have that crucial support when they need it most. These centres give us a frontline view of the problems our customers face, ensuring that we can respond quickly and appropriately. Whether it's coordinating with a hospital, reassuring a panicked parent or guiding the traveller through the unfamiliar, we aim to provide a service that truly supports them in moments of need.

The speed and quality of our response are absolutely critical. Travel insurance is only as good as the support you receive when you need it most. It's during these moments – when a young traveller, often far from home for the first time, is navigating a new environment – that our service must shine. These aren't just logistical challenges; they're deeply emotional experiences for everyone involved and we treat them with the care they deserve.

Online reviews and net promoter score (NPS) are key metrics for us. They allow us to gauge how well we're meeting customer expectations, particularly during claims. It's true that people are often more vocal about bad experiences, but that just makes it even more important for us to exceed expectations during critical moments. When customers take the time to share a positive experience – whether that's feeling supported during a medical crisis or navigating an unexpected trip cancellation – it's a testament to the hard work and dedication of our team.

Ultimately, we're judged by what we deliver when it matters most. Our goal is to ensure that every customer, especially young travellers who might be navigating their first major crisis abroad, feels cared for and supported. Whether it's through timely action, effective communication or just a comforting voice on the other end of the phone, we aim to be the safety net they can rely on when things go wrong.

Stephen: It's fascinating to reflect on how much travel has changed since we were exploring the world in the late 1990s and early 2000s. Back then, the idea of 'live updates' didn't exist. You had to rely on an internet café and a Lonely Planet guidebook, and sometimes you'd turn up to find that the place you were counting on didn't even exist anymore! Do you think the shift towards detailed research and pre-planning has taken away some of the excitement of the unknown? Or is it just a reflection of how travellers – and their risk tolerance – have evolved?

Nick: There are countless stories like yours, where the unpredictability of travel was part of the charm. That feeling of showing up in a new country with just a guidebook and no set plan was exhilarating, but it also came with its own challenges. The brands that pioneered youth travel decades ago laid the groundwork for the way businesses like World Nomads and others operate today. They normalized the idea of exploring the unknown, but they also showed us the importance of making that experience as safe and accessible as possible.

The world has changed significantly over the last 40 years and risk is now at the forefront of many travellers' minds – particularly for young travellers and their parents. Back then, turning up somewhere without a plan might have been an adventure. Today, that same scenario might feel more daunting than exciting. The risk hasn't disappeared, but the perception and acceptance of risk have shifted dramatically.

One key factor is the changing demographic of travellers. We're seeing far more young women travelling independently now compared to past generations, which brings new considerations for catering to the needs of travellers. The presence of tools like smartphones, Global Positioning System (GPS) and instant access to reviews and safety information has undoubtedly reshaped the way people approach travel. A young traveller arriving in a new country today might feel unsure or even scared, but they now have resources in their pocket that provide a level of comfort and control we couldn't have imagined 20 years ago.

At World Nomads, we've adapted to this shift by focusing on travel safety tools and resources. We provide destination-specific safety advice to help travellers prepare for their trips and navigate challenges they might face. This isn't just about selling insurance – it's also about empowering travellers to feel confident and informed, whether they're heading to a bustling city or an off-the-beaten-path destination.

The thrill of travel hasn't disappeared; it's just evolved. While there might be less of the 'turn up and see what happens' approach, today's young travellers have replaced that with the excitement of connecting through social media, researching unique experiences and creating itineraries that reflect their personal values. That sense of discovery is still there, but it's now balanced with tools and resources that reduce uncertainty and make the world feel just a little bit smaller.

Stephen: In Chapter 14, Simon Lucey mentioned planning a trip entirely through ChatGPT. He said the tool created an itinerary not far off what he would've built himself through traditional research – like reading guidebooks, scrolling Google Maps and diving deep into reviews. I joked with him that, while it's efficient, it takes away the joy of discovery, that thrill of finding hidden gems through your own efforts. His response? 'That was ok for our age group, but things have changed.' And that's exactly it – a generational shift in how people plan travel.

For today's youth travellers, safety and transparency are critical. It's about reassuring not just the traveller but also their parents. They need to feel confident that the destination is safe, secure and equipped with accessible healthcare and other essentials. When I travelled, I barely thought about those things. I remember being in Cambodia when the Khmer Rouge was still somewhat active and travelling through Vietnam during a bird flu outbreak. We just did it without a second thought.

But the world has changed. Destinations and brands need to understand that attracting youth travellers requires a fundamentally different approach compared to, say, business or luxury tourists. Safety, accessibility and the assurance of an enriching and secure experience are absolutely key.

QUESTION

Do you have a clear digital strategy ensuring that young people can engage with your brand on mobile and other devices?

Nick: You're spot on, Stephen. Safety and accessibility are paramount for today's youth travellers, but what's often overlooked is how their extended stays impact the local economy. Youth travellers tend to stay longer and spend more as a result, which creates a ripple effect in the tourism ecosystem.

There was a study about eight years ago in Australia that revealed two out of every five tourism dollars were spent by youth travellers.[3] That's a staggering figure, especially when compared to high-end or business tourism. Youth travellers don't just land in the major cities – they explore the broader country. They're out there visiting smaller towns, engaging with local communities and spending money in areas that might not see much from traditional 'fly-and-flop' tourists.

This aligns with what we've always championed at World Nomads. Through programmes like Footprints, our responsible travel manifesto and our travel safety content, we've consistently encouraged travellers to look beyond the typical all-inclusive resorts. Our audience isn't the type to spend a week by the pool with an endless buffet. They're the adventurers – the ones seeking immersive experiences, off-the-beaten-path journeys and deeper connections with the places they visit.

This focus on adventure and authenticity also shapes our industry partnerships. We tend to collaborate with businesses and organizations that share our ethos – those that cater to explorers rather than those looking for a standard, pre-packaged experience. It's not just about where travellers go; it's about how they engage with the destination and the legacy they leave behind.

For destinations looking to attract youth travellers, this is where the opportunity lies. These travellers aren't just visitors; they're economic drivers, cultural connectors and the foundation of long-term tourism growth. It's about creating an environment where they feel safe, supported and inspired to explore everything a destination has to offer.

Stephen: One of the recurring themes of this book is how the value of a youth traveller often goes beyond just their daily spend. While they may not match the high daily outlays of a luxury traveller, their total trip spend often eclipses that of other segments, particularly as their money is dispersed across a wider range of activities and destinations. Do you see long-term value in nurturing loyalty among youth travellers? For instance, do you aim to create a journey where a traveller starts with World Nomads in their twenties and stays loyal to the brand – or moves across different offerings as their travel habits evolve?

Nick: While youth travel is a significant part of our focus, World Nomads isn't limited to younger audiences. Our policies are available to travellers of any age, so it's more about mindset and the style of travel than strict age brackets.

World Nomads is part of a larger organization called nib Group, a leading health fund provider based in Australia. Within nib's family, consumers navigate different life stages – starting as adventurous explorers with a World Nomads single-trip policy, then transitioning to products like Annual Multi-Trip policies or nib private health insurance as they settle down, start families or retire.

From our Millennial customers – who were our first major demographic 20 years ago – we've seen loyalty that extends beyond one-off gap year adventures. We've had policyholders who have been with us for 15 years. They may not be taking long, backpacking-style trips every year, but they are still travelling regularly and we continue to support them as their needs evolve.

As part of our evolution alongside the needs of our customers, we've launched products like the AMT policy in the UK and US markets. The AMT product caters to travellers who may take several shorter trips each year, such as weekend getaways or city breaks, rather than the extended adventures possible on our single trip policy. The AMT policy does have limits on the duration of coverage for each trip, which vary depending on the plan you choose. However, it offers 12 months of coverage, potentially making it a cost-effective option for those who travel regularly. This product particularly resonates with customers aged 30 and older, who may no longer be the same backpackers taking one extended trip per year as they did in their twenties, but still highly value the ability to travel frequently and with flexibility.

For example, my wife and I use low-cost airlines and route maps to find new destinations, taking two or three short trips annually. An AMT policy fits perfectly for this type of traveller. We've seen similar trends in our UK audience, which has guided the development of this product and allowed us to continue serving travellers as their habits change.

So, yes, we absolutely see the long-term value of nurturing loyalty, whether it's through World Nomads or our broader suite of products. The journey might start with a gap year or backpacking trip, but it's about building a relationship that can grow and evolve with the traveller's needs over decades.

Stephen: That's really interesting, because one of the challenges that came up when I was speaking to Sally Cope in Chapter 5 was the difficulty in tracking travellers once they return home. Inherently, if you've worked in youth travel, you know that youth travellers will often return to a destination at different stages of their lives – for business, holidays or to visit friends and family they met during their earlier travels.

But it feels like some brands and destinations don't fully grasp the long-term value of these travellers. Something like travel insurance, though, is perhaps more akin to a bank account or mobile phone contract – once someone has had a good experience, they're likely to stay loyal because it works for them.

I had a great conversation with Sam Willan, who was formerly at Student Universe. He talked about how some airlines are laser-focused on getting young travellers onto their loyalty schemes, starting with the back of the plane, because they know that in 20 years, those same customers could be sitting at the front with frequent flyer miles and the same loyalty cards.

The more destinations and brands understand that long-term value, the more they'll invest in youth travel from the start. But there are still many who don't seem to realize its importance. Is that something you've seen or agree with?

Nick: Absolutely. We've seen this play out countless times, particularly in the Australian market. At consumer or industry events you'd often find yourself on a stand next to a competitor – everyone vying for the same traveller. But the reality is that, as long as that traveller chooses Australia and experiences everything from Cairns to Brisbane, Sydney, Melbourne, Adelaide and Perth, the entire destination wins.

The real competition isn't the operator or service provider on the next stand – it's other destinations, like Latin America or the US, that are vying for the same traveller. This shift in perspective – understanding that the market is broader than your immediate competitors – was a huge evolution for destinations and brands.

Going back to your point about generational travel, it ties into this as well. Our parents' generation pioneered the kangaroo route, travelling overland from London to Australia, and we might have done something similar, albeit with less planning. But the current generation wants to chart their own path. They don't necessarily want to do what Mom and Dad did – they want their own unique experiences.

That's why destinations like Australia aren't just competing with local surf schools or adventure tours; they're competing with the surf scene in Argentina or adventure tourism in Costa Rica.

What we've seen work well is when destinations and industries club together. Instead of fighting over who gets the traveller's dollars once they're in the country, the focus should be on simply getting them into the country in the first place. Once they're there, the economic ripple effect benefits everyone.

This shift – from competing locally to competing globally – has been crucial. It's about thinking bigger and realizing that the real win is bringing travellers into your destination and giving them the experiences that will make them advocates for life.

Stephen: It's clear that the long-term value of youth travel resonates deeply within World Nomads' strategy. From what you're saying, it seems like your organization recognizes not only the loyalty potential of these young travellers but also how their evolving behaviours shape the industry.

I'd love to dive into some of the innovative ways you're adapting to these shifts. You've mentioned launching new services like SmartDelay™ and Air Doctor, which sound designed to appeal directly to younger generations, but also enhancing the overall travel experience. How has the organization balanced innovation with maintaining that tribal brand loyalty you've built over the years?

Innovation is key for youth

Nick: We are very focused on delivering innovative insurance products with additional services that suit the needs of this ever-changing market who are being more accustomed to receiving perks on top of their primary purchase, whether it's travel, hospitality or food and beverage. Collaborating with our underwriter, Collinson Group, has allowed us to introduce extra features, such as Smart Delay™ and World Nomads powered by Air Doctor, which serve as excellent examples of us ensuring our customers are looked after beyond their insurance policy. SmartDelay™

offers access to an airport lounge when flights are delayed beyond an agreed time limit, while Air Doctor connects travellers with medical professionals, in 78 languages worldwide. These aren't just add-ons; they're solutions tailored to enhance the experience for Gen Z, Millennials and even Gen X travellers like myself.

Brand loyalty, though, is a deeper concept. As humans, we're naturally tribal. It's ingrained in how we identify with the things we love, whether it's the football team you support – which, I noticed, you proudly display behind you – or the brands you trust.

At World Nomads, this tribal connection is something we've nurtured through initiatives that go far beyond just selling insurance. About 15 years ago, we launched our scholarship programmes, offering once-in-a-lifetime opportunities for aspiring travel writers, photographers and filmmakers. These programmes are deeply personal, offering participants the chance to work with iconic brands and mentors in their chosen fields.

What's been remarkable is the repeat engagement. We've seen the same individuals apply year after year, showing how much these programmes resonate. They're not just about winning a prize; they're about creating opportunities, shaping passions and sometimes even launching careers.

Three past winners, for example, have gone on to build successful careers in travel writing and photography. That's incredibly rewarding for us as a brand. It solidifies the loyalty we've built, showing that we're more than just a travel insurance company. We're a part of their journey.

Stephen: It's impressive to hear how these initiatives not only build brand loyalty but also give back to the community in such a meaningful way. It reminds me of something we touched on earlier about the vast competition in your space. With so many options available at the click of a button, standing out is not easy.

Do you feel these scholarship programmes and innovative products have given you a distinct edge in the global market?

Nick: Definitely. They've been instrumental. The competitive landscape is immense – you can type 'travel insurance' into Google and be overwhelmed by tens of thousands of results. What we've worked hard to do is differentiate ourselves by offering more than just a transactional product.

Our brand attributes – like the scholarships, the philanthropic work through Footprints and our travel safety content – add layers of trust and connection. These initiatives make us more than an insurance provider; they make us a partner in travellers' experiences.

This approach has also been critical as we've expanded internationally. In addition to our headquarters in Australia, we now have offices in the US, Ireland, the UK and New Zealand, allowing us to speak directly to local markets. Having local people in these offices means we can better understand and respond to the needs of each market, further building that sense of loyalty.

With Gen Z, the challenge is staying ahead of the curve. They're incredibly savvy and less likely to remain loyal without a strong reason. That's why we're exploring partnerships in emerging areas like e-gaming. Gamers travel frequently for events and their audience numbers are staggering. By aligning with trends like this, we are positioning ourselves not just as an insurance provider but also as a brand that understands and supports the evolving interests of younger generations.

Stephen: It's incredible to see how forward-thinking your strategy is. I have to say that I would never have put e-gaming and travel insurance together. The fact that you are tapping into trends like e-gaming shows how well you're adapting to the future of travel. And I think what stands out most is your commitment to being more than just a provider – you're a part of the story for your customers.

In a world where destinations and brands sometimes overlook the long-term value of youth travel, World Nomads seems to have cracked the code for creating lifelong connections. It's inspiring to see how those connections evolve with your customers, no matter where they are on their journey.

One of the themes that keeps emerging from this conversation – and from what you've said, Nick – is that success in the youth travel market isn't just about having the best product. It's about creating an entire experience around that product, understanding the market on a granular level and localizing your approach.

It's fascinating how you have highlighted the differences between markets like Australia and the UK. Although the language is shared, the purchasing behaviours, expectations and cultural contexts are vastly different. What stands out is that World Nomads has recognized these nuances and tailored its strategies accordingly. It feels like many brands could learn a lot from your approach because, as you said, those who fail to adapt risk losing credibility and relevance with young travellers.

Nick: Absolutely, Stephen. When we launched back in 2002, we were just another small player trying to carve out a niche in a very crowded market. At that time, Google AdWords had just entered the Australian market and we embraced it as our primary distribution channel. Being born for the web gave us a natural edge, but we quickly realized that wasn't enough.

We understood early on that partnering with trusted brands in the backpacker and youth travel space was essential. These were established players – brands that had been in the market for 10 or 15 years and already had the trust of their audience. By collaborating with them, we could position ourselves where it mattered: right in the heart of the buying cycle.

What has been really interesting to observe over the past two decades is how the relevance of travel insurance in that buying cycle has evolved. It's always been part of the decision-making process, but its prominence has grown significantly. Today, we are not just a product to check off a list; we are a key consideration early in the travel planning phase.

That shift has not happened by accident. It is the result of consistently delivering value, fostering trust and showing that we understand what youth travellers need. Now, we find ourselves in a position where start-ups and other B2B businesses are approaching us, asking to partner or be featured in our campaigns.

For example, we regularly hear from companies wanting to align with our editorial calendar or collaborate on marketing initiatives. That is a big shift from when we were the ones reaching out, trying to prove ourselves. It's a testament to the work we've done over the years and how we've positioned ourselves as a leading travel insurance provider for youth.

Stephen: That's such a powerful story – moving from being a challenger brand to becoming a trusted leader that others want to collaborate with. What strikes me most is how this success ties back to your ability to stay at the forefront of the buying cycle, not just reacting to trends but shaping them.

It's clear that for any brand aiming to tap into the youth travel market, the key isn't just selling a product. It's about embedding yourself into the journey, understanding the audience and earning their trust through authentic, meaningful engagement. World Nomads seems to have mastered this balance beautifully.

Conclusion

The youth travel market is a powerful engine of economic growth, innovation and cultural exchange. As Nick's insights illustrate, successfully tapping into this demographic requires more than just selling a product or destination – it demands creating an entire ecosystem of trust, authenticity and adaptability.

World Nomads exemplifies how a brand can evolve from a functional service provider to a trusted companion in the traveller's journey. By embedding social responsibility into its business model, investing in initiatives like Footprints and scholarships, and addressing the specific needs of its target audience, World Nomads has built a loyal customer base that spans generations.

For destinations and brands, the key learning is clear: youth travellers are not just customers; they are influencers, advocates and repeat visitors. By understanding their values, leveraging technology and fostering meaningful experiences, the industry can unlock the long-term value of this market and build lasting relationships that benefit all stakeholders.

KEY TAKEAWAYS

1 **Youth travel as a long-term investment:** Youth travellers often have smaller daily budgets but longer stays, resulting in significant economic impact. Brands and destinations must recognize the potential for long-term loyalty as young travellers return for business, family visits or other purposes later in life.

2 **Safety and trust are paramount:** Today's youth travellers prioritize safety and transparency, seeking assurance not only for themselves but also for their families. Travel insurance has shifted from being an afterthought to a top-tier consideration, driven by global events like Covid-19.

3 **Authenticity resonates:** Young travellers are highly discerning and research-focused. They expect brands and destinations to deliver on their promises, particularly around sustainability and social impact. Initiatives like Footprints demonstrate how transparent, authentic efforts can create lasting connections with this audience.

4 **The role of digital engagement:** Peer-to-peer recommendations, online reviews and influencers play a critical role in shaping perceptions and decisions. Brands must actively engage in these digital conversations and deliver exceptional service to maintain positive reputations.

5   **The power of partnerships:** Collaborating with established brands and local organizations helps build credibility and reach in the youth travel market. Aligning with businesses that share similar values amplifies impact and creates cohesive travel ecosystems.

6   **Adapting to trends and technology:** From annual multi-trip insurance policies to partnerships in emerging fields like e-gaming, innovation is essential to staying relevant. Understanding generational shifts and evolving travel behaviours ensures that products and services remain aligned with market needs.

Notes

1   K Ian. Travel, insurance and personal responsibility: When things go wrong in an unpredictable world, *Guardian*, 28 August 2022. www.theguardian.com/business/2022/aug/29/travel-insurance-and-personal-responsibility-when-things-go-wrong-in-an-unpredictable-world (archived at https://perma.cc/N7D2-NU7A)

2   L Graham. A quarter of Brits don't know what to do when sick abroad, World Nomads, 12 November 2024. www.worldnomads.com/in-the-news/media-releases/medical-assistance-while-abroad (archived at https://perma.cc/FE7F-G7LJ)

3   Tourism Australia. The youth sector, Tourism Australia, nd. www.tourism.australia.com/en/about/industry-sectors/youth.html (archived at https://perma.cc/X3KT-WEJF)

7

Challenges facing the youth travel sector and what they mean for the tourism industry

In this chapter we will explore the state of play of the industry in the UK and what is holding it back from being the best it can be, with views from Emma English, the Executive Director of the British Educational Travel Association (BETA). Reading through this chapter, you should really consider whether your destination or brand are putting up the same barriers, or, alternatively, whether you are really looking at the long-term value of what these young people bring to where you operate. As you will see from some of the statistics in this chapter, the barriers put up are often not from the private sector, but are a result of governmental decisions that have an unintended impact on many different private businesses.

BETA and its role in youth travel

Stephen: What is BETA, and what role does it play in the UK's youth travel sector?

Emma: BETA is a not-for-profit, membership trade body and we exist to raise the profile of youth and student and educational travel to, from and within the UK. We also aim to help to create commercial opportunities between buyers and suppliers within the sector. Our members are the entire supply chain from all forms of accommodation, be that hostels, homestays, student residences, transport from airlines to rail to car, ferries and coaches. Of course, we also have a different selection of educators and language schools. Then all those other sorts of supplier services that go alongside making an excellent itinerary for a young person, such as tickets, tours, food and beverage, and even professional services like insurance and marketing. We really help by covering the entire supply chain of the sector.

Stephen: This shows how varied youth travel is, where probably the perception of youth travel within much of the travel sector and among people in other sectors is that it is just university, just a school trip or a backpacker. This shows it is actually all and everything, and more, with the industry having products and services that are age-related to the youth travel sector rather than a particular purpose.

Emma: It's not just people taking part in a school trip. You can head to the Tower of London today and see hundreds of young people in groups lined up in their hi-viz vests from both UK and international schools, but it's far broader than that. Our sector delivers group travel, junior groups and cultural groups, but we also have providers focused on individuals, for example work experience operators including volunteering and gap year experiences. It is a pretty vast cross section of organizations.

TIP

This highlights the huge variety within the youth travel sector both in terms of reason for travel and the different types of engagement with a destination and what it has to offer.

Founding BETA: A unified voice for the sector

Stephen: Obviously, associations are often created because of a need and you have been with the association since it was founded 21 years ago. Why was BETA founded and what was its purpose?

Emma: The association was initially founded through the attendance of organizations at an annual international conference and it was ultimately because the UK delegation really felt that they needed a central voice representing them. They needed to come together to create a body in the UK because they realized that lobbying at a global scale is just not achievable nor effective. So that idea turned into forming a national body to represent the needs of businesses in the youth, student and educational industry. All credit to those initial UK stakeholders because they really came together, put their money where their mouths were and said, 'Let's start this organization.' They obviously really believed in youth travel, not just in the short term but also the long-term value of it.

To start with we had 10 founding organizations. We had no funding from external sources, government or otherwise, so everything was supported by those original founders to enable us to set up and establish the association.

Looking back at it, it was quite remarkable that we were able to pull together the association in such a short amount of time and with such a small founding group of people. One thing that the UK does have is a lot of trade bodies and associations, not just in travel but across sectors. However, there was no trade body that just had the interests of youth and student and educational travel, and 20 years ago it was very much misunderstood. It was viewed as the poor sub-sector of the main travel and tourism industry. It was perceived by many as a low-yield business and less important to the general sector. Our goal and ambition was to make sure that people understood it and its long-term value to the UK. We wanted people to realize that this is the start of all travel and tourism experiences.

We also wanted to share that getting young people out and about, making them better global citizens and improving their language skills is vital to the UK. The fact is that those experiences ranged from school travel right the way through to working and living and travelling overseas. While it was very important to have that support, we quickly found ourselves in the position of bringing together a whole series of organizations, from small to large, with direct or indirect interest in the sector. We had to start building our network and very quickly. By the time we launched at the Houses of Parliament in May 2003 we had a huge amount of support and our membership was building at a good pace, so it was obvious that there was a real need for this type of organization to raise the profile and speak in a unified voice on behalf of the industry.

TIP

The importance of creating a trade body representing your niche cannot be underestimated. Whether it is to share best practice or statistics, or to lobby government for changes to policy to improve the sector, trade bodies are so important.

Quantifying the sector: The power of data

Stephen: May 2003 is a long time ago. I was about to graduate from university with dreams of setting out on my around-the-world adventure. Now, being involved in the sector, it's been amazing to realize and see how broad it is, but also how important having it is to have a central focal point to get everyone together and discuss the positives and negatives of the sector.

One of the things that I have experienced, and I know you have too, is that many people think that youth, student and educational travel is a niche sector, but actually it isn't. Could you explain the size and scale of the sector in the UK so the readers can understand the context of this?

Emma: Creating data is one of the crucial things that we had to do, as it really was not quantified or understood when we set out to create the association. We realized if we wanted to get any traction and understanding of the sector in terms of lobbying, we needed to know its volume and value in order for both the government and also the wider industry, to understand we were not 'niche'. The data and industry insights would allow us to gain PR and help us with positioning the sector as a very important part of the wider tourism ecosystem. Therefore, we set out on a data collection project using our membership, and wider national and international data sets to allow us to benchmark the UK. We are just working on a new set of data at the moment, which we'll publish later this year, but I can share the data from the last study.

In 2023 there were 12.5 million young visitors who came to the UK and that accounted for 33 per cent of all international arrivals inbound into the UK; that was valued at £34.6 billion worth of revenue for the UK economy. If you think that a third of all international arrivals were young people, that is significant and definitely not niche. There are also global figures which show that it's estimated that one in four international travellers is a young person.

As you can imagine, we have managed to create hundreds of pages of data, but we have been clever in the way we share this, as we want to make sure that we are aiming the data at understanding the scale of the sector. For example, we showed that the volume of young people coming to the UK was enough to fill Wembley Stadium 166 times, or for example 100,000 airplanes would be needed to deliver that amount of young people to the UK. We found this helped them better relate to and understand the enormous opportunity that this sector has, if supported correctly.

The UK government's role: Understanding and missteps

Stephen: Any business getting revenue from a market that was 25 to 35 per cent of your total revenue would not call it niche. What is interesting to see is where different countries have different views on how important youth travel is towards their tourism industry and greater economy. With that in mind, over the last 20 years you have obviously had a lot of dealings with the UK government, spanning different parties and varied political ideals. Do you think they have understood the longer-term value of youth, student and educational travel, and has that understanding ever really changed over the last 20 years?

Emma: We have certainly made some inroads and I think the answer is mixed. Have they understood the enormous potential that this sector is able to deliver? Possibly. Do they do enough to help us to achieve those things? No.

There is very little global marketing for this industry, if we compare it to other countries that truly value it as part of their economic growth. The UK just does not really do that. In fact, what various UK governments have done is, time after time, make it more difficult for young people to come to the UK; whether through changes in visa requirements or increasing the fees. For example, the UK has increased visa fees by almost 20 per cent in the last year alone. All of those things make us less attractive as a destination for young people.

Do they understand it when we deliver them our statistics? I think so, but unfortunately, what we lack also in the UK is that cross-departmental collaboration and understanding, so decisions are made that sometimes cause damaging, though unintended, consequences.

To give you an idea about how it works, the travel part of our membership sits under the Department for Culture, Media and Sport. If we are talking about the education part of the sector, we sit with the Department for Business and Trade or the Department for Education. When we have certain issues around immigration, travel restrictions, removal of certain criteria, which we have seen since Brexit, then we have to deal with the Home Office or the Foreign Office.

With so many issues and opportunities for the UK sitting in so many departments, it is frustrating that there's not enough of that cross-departmental understanding to make decisions that join up. If that was to be corrected, it would really help us to propel forward and create many more opportunities. I am pretty sure that people are making decisions in the Home Office about things that affect our sector without any real understanding of some of the consequences, and that comes down to the understanding of the variety and enormity of the supply chain and the impact it will have on the wider UK economy. We saw that in Covid with the support that wasn't provided, for example, to businesses in our sector because they were not retail travel agents with a shop front, despite delivering millions of pounds of revenue to the UK economy through providing services to the youth target market. So, there are still lots of challenges, but we have built up, and particularly since Covid we've built up some good support across various government departments. We have been working hard with some of the experienced peers who really care for the sector to help and support and we now have people speaking on behalf of the sector, which in turn has created a lot more understanding across the political spectrum of what we are doing and trying to achieve.

It is just constant activity that we need to do, which is why things like our data and our research is so important to have. It allows us to keep going back, we can provide the anecdotal evidence, we can help them to understand. We explain to MPs about young people travelling and they respond with, 'Oh well, my daughter, friend, sister, mother taught in France when she was 19 on a gap year', and it then starts to reinforce our argument.

From a British perspective, why would we stop that? Some of the rules since Brexit have made it hard for school-age Europeans to travel here. Under-18s on an organized school group trip who have got parents waiting at home for them and are fully escorted are not going to escape and work in the black economy. We should make these activities easier because travel from a young age is so important.

Do they understand? Yes and no. There's still an awful lot of work that we need to do. But you know, we are consistently chipping away at it. And I think since Covid, more people now know that young people cannot travel on ID cards and why it is such a huge problem. Government officials and other parts of the tourism industry now see these blockages and end up filtering out into the other sectors, which is important as that means the rest of the tourism sector have started sharing our messages. They know the family sector and luxury sectors will be impacted if this type of travel is restricted. If we allow these school-aged children and their teachers to go elsewhere for these experiences, other loyalties are made. It's a bit like the Australian ancestry travel that happens in the UK. This is where young people retrace their family tree by travelling across the British Isles. You know that in the past your parents did that journey and so will you. But if we stop young people coming here they'll choose other destinations and that cycle will continue elsewhere. These are people who do not need big-bucks advertising; they will come back via word of mouth.

TIP

This shows that even established destination such as the UK, do not always grasp the full impact of the power of youth travel. Instead, the sector gets drawn into larger political policies creating unintended consequences.

Global comparisons: Learning from the best

Stephen: One of the things that we're exploring in this book is the long-term value of attracting young people to your destination and I use my own personal example of backpacking and ending up setting up a hostel in London from being based in Sydney. Now if I had not gone to Sydney, I would not have set up the hostel and I probably would not be here now.

There are some destinations that understand the full long-term value. On top of the statistics that are known about the value of this sort of business, it doesn't even include what the indirect value is in the future.

Those people who are company decision-makers, or where and when you personally invest money into stocks and shares, you have an outlook of whether it is going to be a five-year investment or a 20-year investment. What is needed, by the sounds of it, is if you are investing in youth travel it's a 20-year investment. In 20 years, you will get that back and more because obviously these travellers are most likely to come back without the need for marketing to them again. I think what is interesting is where people are just looking for these short-term wins in tourism, whether it is politically or financially driven. I think for some reason, here in the UK, I feel like youth student and educational travel seems to be the bit that gets targeted when spending cuts kick in when it comes to marketing, when actually it is the better longer-term investment.

Emma, you deal with a lot of associations and companies based all over the world. Who do you think does it best and why?

Emma: I think the Australians are the number one in terms of their understanding of the market, its value, why it's important to their economy and to their jobs market. They have always understood it deeply and I think that opinion has never changed since we started 20-odd years ago. Even in the last couple of years, with the reciprocal changes to youth and mobility schemes, together we opened the age range between the UK and Australia so that it was expanded from 30 to 35 years old as an upper age limit. Straight away, the Australians understood that meant they could capture a much larger audience; approximately an extra 6 million young people would be eligible to travel on a Youth Mobility Scheme (YMS) visa. That was on top of the approximately 11 million 18- to 30-year-olds already eligible to live, work and travel in Australia. Since they are much more invested in youth in their tourism brands strategy, they quickly installed billboards in Trafalgar Square saying 'Come visit us down under', targeting young people up to the age of 35. No surprise that their tourism stats in young people from the UK grew back quickly.

Sadly, in response, the UK tourism boards have not invested in attracting young Australians here. We have got that same reciprocal arrangement, but our tourism boards are not targeting that audience.

Australia as a tourism destination and brand is very agile with these campaigns as things move on within the sector. They've always done it very well. I think also more recently the Canadians have also done very well and made some very great moves in terms of attracting young people to live, study and work.

Post-Covid in the UK we were having a labour market crisis from a hangover from Brexit and career changes through Covid, which meant that the talent pool for industries like hospitality had shrunk. This meant that many businesses were struggling with simply reopening due to a lack of people to operate their businesses and in turn impact the country's economy. Obviously, that was a global issue as well, but the Canadians were very nimble with their government's support with their policies and they quickly put in visa policies to allow young students to be able to work and increase their hours, which filled that gap in the labour market. This really helped to address these issues and in turn helped the economy get back on its feet post-Covid. It was initially treated at that point as a pilot scheme, but they understood that, for their economy to grow and/or recover, first of all they needed people back in the workforce. They needed the businesses to be open, so they really made that happen through international students or international students that were already there but not engaged in work.

It is obvious that governments can do that if they want to, and that was demonstrated in this example.

Stephen: As you said, the example of the billboard in London and nothing in Sydney is a prime example that a destination cannot be lazy. They must market consistently and market to youth specifically. The next question is, how you have seen the industry change? Because in 2003 there was no such thing as an iPhone, Facebook didn't exist, TikTok wasn't even a thought. I assume youth travel and marketing to that target market must have changed at an overwhelming pace over the last 20 years. What are the biggest changes that you've seen?

Emma: Well, as you say, social media is everything to our audience and has really allowed it to propel and grow some of the brands in our industry quickly because young people are digital natives and early adopters of technology, and they are the young influencers. Having social media really enabled our sector to grow, so that was really big, and we have seen some of the businesses that started with BETA 20 years ago that were small single operators grow into now large pan-UK and pan-European multi chain operators, which has been amazing to see. There has been some real growth and, for a time, youth travel was the fastest growing travel segment, and because of that an awful lot has changed over the years. Included in that is that backpacker-style hostels were the 'hot' investment when it came to hospitality real estate in Europe and Australasia between 2008 and 2018 with large brands like Accor joining the sector.

Entrepreneurship in the sector

Stephen: One of the things that I found interesting is that a lot of people working in youth travel have had some sort of youth experience as they were growing up. Is that something you think that has been driven by the fact that 30 years ago flights and the ability to travel became more accessible globally? Do you think that has increased, given it has grown as a sector, and therefore more and more young people are now experiencing youth travel? People say, 'I did a language school trip in France, I backpacked around the world, or I studied abroad.' Do you find that you've seen an even greater number of professionals in the sector being former or current youth travellers?

Emma: Absolutely – you should think of youth travel as an incubator for entrepreneurship. So many of our members are people that have experienced some sort of travel like you when you went backpacking in Australia and came back with a hostel brand and ideas of how to expand other businesses. We have seen that so frequently within our membership, people that have had these experiences in gap years or volunteering, in study abroad and have sort of gone, 'Okay, this is great, but I can do it better', and it's really inspired them to come back and be a business owner and an industry leader and young entrepreneur. It really is an incubator for that style of exciting startup travel business.

Stephen: Going back to the title of the book about this long-term value, this is another indirect value that youth, student and educational travel brings. That you may go home with amazing ideas from your travels. You may have also advertised for a young person to come to your country or use your brand, but when they go home the inspiration of the travel inspires them to create something amazing in their home country, but often trading with where they have travelled and had the best experiences.

Emma: It is all about the welcome. BETA had undertaken several consumer-based studies in the past when we asked young people in the UK that are in a backpacker's hostel or in a language school setting, 'Do you intend to return to the UK after your trip?' The answer was always 'Yes' and an enormous percentage said 'Yes, within the next five years.' I think that is why we also need to try to demonstrate the long-term value as these young people are the repeat visitors over a lifetime for the UK. If you can get the welcome and experiences right, it has so much more potential.

Conclusion

One can really see in the case of the UK that youth travel is a sector that is vital to the total ecosystem of tourism, but without the spend, support or recognition that you would think for such a large part of the sector. What Emma highlights is that, despite challenges from a governmental level, if the sector works together amazing things can happen to ensure the success of youth travel, but also in turn create huge amounts of soft power and success for a destination. Emma also highlights that the benefits are not only returning visitors from a leisure tourism perspective but also the entrepreneurial impact of young people travelling to and from your destination. Although some of the soft power impact of this type of travel is somewhat hard to quantify, it cannot be underestimated as a vital part of the importance of youth travel in the short-term revenue but also, most importantly, the long-term impact of the sector on a destination. I thought it really interesting to hear how an established and well-known destination like the UK feels in many ways that young people will come here regardless, but in reality the competition globally is growing, and at some point in the future, unless things improve with regards to the welcoming nature of the UK, then young people will start choosing elsewhere for their short- or long-term adventures.

KEY TAKEAWAYS

1 **Youth travel's economic importance:** Youth, student and educational travel represent a significant portion of the tourism industry, contributing immensely to the economy. In the UK, youth travel accounted for 33 per cent of all international arrivals in 2023, translating to £34.6 billion in revenue. This underscores that youth travel is not a niche sector but a critical part of the tourism ecosystem.

2 **Comprehensive sector representation:** BETA plays a pivotal role in advocating for the youth travel sector. It represents a wide array of businesses, including accommodation providers, transport operators, educators and support services, ensuring that the entire supply chain is supported.

3 **Government challenges and advocacy:** Despite the sector's substantial contributions, governmental policies and visa restrictions often create obstacles for youth travel in the UK. These include increased visa fees and stringent entry requirements post-Brexit. BETA's work highlights the need for cross-departmental collaboration and unified governmental support to leverage the full potential of youth travel.

4 **Global best practices:** Countries like Australia and Canada are exemplars in recognizing and capitalizing on the long-term value of youth travel. They implement youth-friendly visa policies, strategic marketing campaigns and innovative measures to attract young travellers, which significantly boost their economies and address workforce shortages.

5 **Long-term benefits and investment:** Investing in youth travel is a long-term strategy with substantial returns. Youth travellers often form strong emotional and cultural connections with destinations, leading to

repeat visits and sustained economic benefits. Additionally, youth travel fosters entrepreneurship, as many young travellers return home inspired to create new businesses or return as investors and leaders in their fields.

6   **Marketing and technological shifts:** The rise of social media and digital platforms has transformed how youth travel is marketed. Young people are early adopters of technology, and engaging with them requires strategies that resonate with their digital habits and influence-driven decisions.

7   **Entrepreneurial incubation:** Youth travel acts as an incubator for entrepreneurship. Many industry leaders and innovators started their journeys as young travellers, drawing inspiration from their experiences to establish successful businesses and ventures in the travel and tourism sector.

8   **Barriers and risks:** The UK risks losing its competitive edge if it fails to address the barriers hindering youth travel. Countries that proactively create a welcoming environment for young travellers gain long-term soft power and economic benefits that extend beyond immediate revenue.

9   **Unified voice for change:** For youth travel to thrive, collective action is essential. Industry bodies, private businesses and government stakeholders need to align efforts to advocate for supportive policies and investment in youth travel as a valuable asset to the nation's future.

10   **Soft power and cultural influence:** Youth travel not only boosts economic growth but also enhances a destination's soft power. The connections and experiences formed by young travellers can lead to lifelong advocacy for the destination, enriching its global reputation and influence.

8

Maintaining brand loyalty

Transitioning youth travellers to different brand offerings

In the hospitality industry, the strategic pursuit of customer loyalty is more than just an exercise in marketing – it's the cornerstone of sustainable growth for all big international hotel brands. In this chapter, we dive deep into an insightful conversation with Ed Purnell, Vice President of Hotel Marketing for EMEA at Marriott International, to explore the many layers of brand building, customer engagement and market strategy that drive success across a global portfolio. Ed's extensive experience in the field offers a unique vantage point on how one of the world's most prominent hotel chains adapts to evolving consumer behaviours, particularly among younger demographics.

Stephen: Could you just explain your role and your main responsibilities at Marriott?

Ed: I have been with Marriott since 2017, but my experience in the industry extends well beyond that. My primary goal is to ensure that the hotels that have chosen to partner with Marriott – whether under a managed or franchised model – are maximizing the benefits of that relationship. This involves helping them optimize their marketing and distribution strategies to remain competitive within their local context, while also leveraging the extensive marketing and distribution capabilities that a global organization like Marriott can provide.

   I oversee operations across the Europe, the Middle East and Africa region, which includes everything from the UK to the Middle East and extends down to South Africa. Currently, this encompasses about 1,100 hotels, representing around 90 per cent of our 32 brands.

Given the diversity of this area – characterized by a multitude of languages and varying customer behaviours – it's crucial for us to tailor our approaches accordingly. In addition to the overarching responsibilities of my role, I am committed to ensuring that we implement strategies that effectively capitalize on customer behaviour throughout the region.

Attracting younger demographics through targeted brands

Stephen: Marriott has developed a vast array of brands over the past two decades, particularly with the addition of new offerings. Has there been a specific emphasis on attracting a younger demographic, particularly those under 35, commonly referred to as the youth travel sector? Additionally, are there particular brands within Marriott that cater more specifically to younger travellers?

Ed: Unquestionably, certain brands within Marriott appeal more to younger travellers for various reasons, including the style of stay, affordability and the overall aspiration associated with those brands. We like to think of it as having a 'fit for every preference'.

When we analyse our brand archetypes, it's often straightforward to identify the target demographic we envision for each hotel. However, we are frequently and pleasantly surprised by the diverse range of guests that ultimately stay with us. This prompts us to delve into the data to understand the different entry points for guests into our brands, leading to insights about potential development opportunities. We can then adjust our marketing strategies – either amplifying or scaling back – based on the age demographics or customer segments engaging with specific brands.

Given the multitude of brands we offer, we are certainly not alone in this approach; competitors like Hilton, Accor and IHG also maintain extensive multi-branded portfolios. The goal for us is to unify this range of offerings under a consistent framework, which is where Marriott Bonvoy comes into play. Marriott Bonvoy is our loyalty programme, serving as a tool to bring all our brands together. Within the Marriott International portfolio, several brands demonstrate stronger engagement with specific demographics and we definitely consider age-related marketing strategies when targeting different age groups.

> **TIP**
>
> When you have a number of brands, deep dive into what markets and demographics could be most engaged with specific brands. This is the most efficient and effective way of marketing with the current tools available to businesses.

Leveraging loyalty programmes for lifetime value

Stephen: In some of my other conversations, particularly regarding destinations, a common challenge is the inability to track the customer journey comprehensively. However, with a marketing tool like Marriott Bonvoy, you have the opportunity to see where someone engages with the brand and how their journey unfolds. Can you observe whether individuals who connect with the brand at a younger age tend to remain loyal over time?

Additionally, we've discussed the effectiveness of cross-selling strategies aimed at encouraging guests to upgrade their experiences from budget accommodation to luxury offerings. For instance, someone might start with a weekend getaway in Europe or New York and, years later, celebrate a honeymoon by staying at a luxury JW Marriott. How does this progression play out within your brand strategy?

Ed: Without a doubt, loyalty marketing centres on customer lifetime value. The sooner you can guide a customer along that journey, the more valuable their lifetime relationship with you will be. Interestingly, you could argue that the loyalty aspect begins even with older generations – parents or relatives of young travellers. They may have visited a resort or destination within the Marriott portfolio during their childhood, without being the ones making the financial decisions.

While some children enjoy making holiday choices for their families, they are often not involved in the actual transactions. However, if they had a rewarding experience – whether through a hotel activation geared towards families, food and beverage offerings, or the overall atmosphere of the hotel – there's a good chance that when they finally have the opportunity to make independent travel decisions as adults, they will choose Marriott. That scenario is ideal for a marketer; the goal is to engage them on their journey as early as possible.

I believe it's crucial to forge brand relationships at a young age. This connection can stem from individual decisions or arise from a situation where someone works for a company that has contracts with Marriott, serving as a catalyst for their loyalty journey.

From a marketing perspective, there are significant opportunities for us as an organization to attract customers. This principle applies universally in consumerism – whether in hotels, banking or fast-food restaurants – the goal is to build long-lasting relationships and provide solutions for various reasons for travel. Additionally, this could extend to non-travel-related services such as weddings, dining or golf.

Even if the primary offering isn't a stay, having that trust and advocacy from the end user means that, when they need a venue for a meeting or an event, their first thought might be, 'I wonder if that would be a good place.' Their prior positive experiences encourage them to choose us again.

TIP

This shows that lifetime value is important to the biggest brands in hospitality and should be to destinations, too. Loyalty schemes like Bonvoy are tools that allow brands to get engaged early in a traveller's career and then move them through the brand portfolio as they get older and travel for different reasons.

Stephen: There's an interesting theme regarding how individuals connect with a brand. I've had conversations where children express a desire to embark on the same journey their parents did – particularly Australians travelling to the UK. Those journeys in the seventies, eighties and nineties have left a lasting impression, and now kids are saying, 'Wow! I want to have the same adventure my parents experienced.'

This sentiment is likely mirrored among Americans as well, as there was a significant movement in the seventies and eighties for young people to travel to Europe and have the adventure of a lifetime.

Ed: We have a significant number of Americans who are eager to explore their genealogy. When they open their Marriott Bonvoy app and begin searching for destinations like Scotland, Ireland, Italy, Greece or Poland, they can continue their journey of loyalty in a way that holds deep personal significance for them.

This capability enriches their travel experience by aligning with their motivations for visiting these places. A major part of our business is ensuring that we provide safety, security and memorable experiences that resonate with their personal rationale for travel.

Stephen: Another interesting trend that has emerged is that children are increasingly becoming influencers for their parents when it comes to travel destinations. For instance, a child might express a desire to go to Greece, and the parent insists on staying at a Marriott. This creates a unique challenge for marketers like yourself, as it's essential to ensure that the hotel is relevant to the younger traveller while also meeting the preferences of the parent.

Ultimately, it's a balancing act that depends on the types of brands you offer, as well as the source market and final destination. Finding the right combination to attract both demographics is crucial for a successful marketing strategy.

Ed: For large organizations like ours, as well as competitors like Hilton and IHG, having a significant global footprint allows us to legitimately offer a wide range of choices in major markets such as London, Dubai, Paris, Rome and Barcelona. This should create a win–win situation for everyone involved.

While children may have specific interests – such as wanting to visit a city because Lionel Messi is playing there or Taylor Swift is performing – the parents often prioritize immersing themselves in the rich culture of the destination. They appreciate the ability to accommodate their child's preferences while also fulfilling their own desire for cultural experiences. The greater the footprint these brands can establish across the globe, the better it is for all parties.

Engaging customers beyond stays

Stephen: You mentioned Taylor Swift, and it's worth noting the introduction of Bonvoy Experiences. Unlike traditional loyalty schemes that primarily reward points redeemable for hotel stays or food and beverage within the hotel, Bonvoy Experiences offer a much broader array of options. These experiences cater to a wide demographic and diverse interests, including those that appeal more to younger markets.

Is this strategic approach intentional, aiming to position Bonvoy as a more youthful brand offering rather than solely targeting traditional corporate or luxury travellers?

Ed: Absolutely. We need to challenge the status quo regarding the reasons for travelling. When it comes to redeeming points you've accrued over time, we have adopted a multifaceted approach. First and foremost, we want guests to consider the various redemption opportunities within the hotel space. We encourage them to reward themselves for the points earned through corporate travel by enjoying leisure stays.

However, we also aim to extend beyond traditional hotel stays by offering curated experiences that cater to different interests – whether in sports, food and beverage, culture, or music. Our dedicated team is constantly exploring ways to create memorable experiences that enhance a guest's love and affinity for Marriott Bonvoy.

While we curate experiences that connect to the essence of travel and the reasons people travel, we also recognize that different offerings resonate with different individuals. For example, partnerships with icons like Taylor Swift, Manchester United or Formula One allow us to align our brand with certain archetypes, similar to our brand marketing efforts.

This approach extends to providing compelling reasons for redeeming points and deepening the relationship with Marriott Bonvoy. It's been an exciting evolution; we launched Marriott Bonvoy in 2019 when, after Marriott acquired Starwood, the three loyalty programmes – Ritz-Carlton, Starwood and Marriott Rewards – were combined. Each year, we strive to stay fresh and innovative, offering new and exciting ways for guests to embark on this journey of lifetime value with us. This strategy is just as relevant for younger travellers as it is for Baby Boomers or those attracted to our classic luxury brands due to their preconceived notions of what to expect.

Stephen: I find it interesting how my perception of Marriott has evolved over the years, particularly in comparison to what I thought a decade ago. The transformation brought about by Marriott Bonvoy and the alignment with various brands, especially the new experiences on offer, has significantly impacted my impression.

The introduction of younger brands like W and Aloft through the acquisition of Starwood has certainly contributed to this shift. It feels like Marriott has actively sought to innovate rather than relying on the same old strategies. This stands in contrast to many airlines, where loyalty schemes have often remained static in their approach. While some airlines have begun to experiment with different strategies, such as allowing customers to bid for upgrades or gamifying the loyalty experience, many still stick to traditional models. Marriott's willingness to adapt and explore new avenues is commendable and likely plays a crucial role in enhancing brand loyalty.

Ed: When we launched Marriott Bonvoy, it was during a digital era, or more specifically a mobile-first era. We quickly focused on developing our mobile app, which naturally appeals to a younger audience. Interestingly, we've also had to educate older generations on how to navigate the user experience of the app. Once they get accustomed to it, they realize how simple and effortless it can be to book a hotel or research their stay.

Currently, mobile is our primary distribution channel, and while there's still much more our app can achieve it has already surpassed many of our previous distribution methods over the decades.

As we continue to digitally transform our organization in the coming years, we will look to companies that have disrupted traditional business models – whether it's Amazon or Airbnb. These organizations provide us with valuable inspiration on how to enhance our storytelling, marketing and distribution strategies. This includes not only the hotel stay experience, but also transformational experiences associated with that stay, many of which can further enrich the concept of loyalty.

TIP

If you are not mobile-ready as a brand or destination, you will struggle to capture the full potential of travellers who may be interested in engaging with you. Any strategy going forwards, especially aimed at younger demographics, should be 'mobile-first'.

Social media and the rise of TikTok

Stephen: I have another question regarding your role. While there's undoubtedly a wealth of inspiration to draw from other hotel brands, I'm curious about where you seek inspiration beyond the hospitality industry. When it comes to developing strategic or campaign-based ideas, are there specific sectors or examples that have recently inspired you, particularly those aimed at younger audiences?

Ed: We always take into account our brand identities and, given our global presence, it's crucial to align these identities with the specific location and destination. This is a key responsibility for my team and me. We work to ensure that the essence of the brand resonates within each hotel while also reflecting local trends and touches.

In today's 'Instagrammable', social media-driven era, it's essential for people to showcase their experiences and demonstrate where they are at all times. They love sharing visually appealing moments with their friends, followers and communities, saying, 'This is where I am – you'd really love it!' The power of influence and advocacy is significant.

We aim to incorporate this into much of what we do. At the hotel level, we encourage connections through linking strategies and hashtag initiatives to foster user-generated content that enhances our storytelling. This is particularly effective with brands like Moxy, Aloft and W Hotels, which naturally attract a socially savvy audience. However, we also see this trend gaining traction within our more traditional brands, as social media's impact has become pervasive.

For instance, we've recently launched a campaign with Sheraton centred around the theme of gatherings, in partnership with Reese Witherspoon's book club. This initiative targets a broad audience, appealing to those who share a passion for reading and personal growth. It aligns perfectly with our vision of Sheraton as a welcoming space for gatherings, regardless of age or background. By leveraging social media we can bring together diverse demographics, and much of this engagement is driven by younger audiences.

Stephen: Regarding channels, what emphasis does Marriott place on social media, particularly platforms like TikTok and Instagram? Is the focus primarily on creating experience-led content aimed at younger audiences, such as Millennials or Gen Z? Alternatively, are other channels used more for hotel-level engagement while brand-building efforts for Marriott occur through different means?

Ed: We aim to match specific social channels with particular brands. Our approach emphasizes engaging on social media at both the brand and portfolio levels. Hotels are responsible for their own social media engagement, especially when it comes to interacting with guests – both prospective and past. This includes proactive outreach where hotels respond to conversations about local attractions or great experiences, as well as reposting, resharing and thanking people for their comments.

In contrast, proactive social media strategies fall under the domain of brand and portfolio management. We invest significantly in marketing and content associated with this strategy, ensuring that we highlight stories or content that are 'TikTok-worthy'.

We align this with our brand strategy because we want to maintain control over the messaging. For instance, it wouldn't be appropriate for a high-end hotel targeting an older clientele or a B2B audience to utilize TikTok. Instead, we advocate for each hotel to be mindful of social media opportunities based on their guest archetypes rather than jumping on trends for the sake of it. Our approach must be strategic.

Stephen: One recurring theme in my conversations with various individuals is the belief that TikTok will soon become a primary platform for travel searches, potentially surpassing Google. Have you observed any trends that support this idea? For hotel brands, is Google still the main driver for travel-related searches, or are we beginning to see a shift based on specific brands and the demographics they target?

Ed: Personally, I always prioritize data because I need to be convinced that our marketing investments – whether in dollars, time, or energy – are truly effective. I would never claim to be knowledgeable enough to definitively answer your question. Instead, I rely heavily on our agencies and on individuals who are closer to the younger demographics and engage with these platforms daily. I trust their insights.

Unfortunately, I was born into an era where these shifts came to me later in life, so I need to be convinced about investing in social tactics. However, the key for someone like me is to surround myself with people who understand these trends and can build a compelling strategic case for investing time and resources in certain projects.

Regarding your question about whether we might see a shift in spending from Google to social media channels, I believe we need to optimize and maximize both. If anything, we need to push the hospitality industry away from its traditional practices, which still dominate how many people book their travel. Many continue to rely on outdated habits.

My primary responsibility is to encourage hotels to examine their data and make informed decisions about spending rather than sticking to methods simply because that's how they've always operated. This transformation is a marathon and it's far from over. Each year, during budgeting discussions, we challenge hotels to reflect on why they continue certain practices.

I'm not opposed to trade shows; there's a time and place for them. However, the significant funds hotels allocate to attending these events often stem from a reluctance to change. I advocate for reallocating those resources into digital and social tactics.

We need to have conversations at the hotel level about investing in both digital and social marketing, ensuring they complement one another instead of relying on traditional marketing methods that lack demonstrable returns.

TIP

Use statistics to choose your channels and level of spend. Be nimble as things are changing quickly and you need to be adaptable, no matter how big or small you are as an organization.

Strategic investment in digital transformation

Stephen: It's quite interesting that hotels have often lagged behind other industries, particularly retail, which has led the way with instant purchasing and seamless collection. Everything is very interconnected now, including live availability. Yet, even in hotels, we sometimes face delays, like a half-hour lag in updates.

You mentioned relying on experts for insights into different channels, which is crucial for deciding where to invest. Employing younger individuals to drive youth-oriented marketing is another smart move. This is an area where some destinations struggle, while others, like Australia, excel at intelligent marketing. They effectively empower young people to become brand advocates, whether directly or indirectly.

In many ways, hotel brands of your scale need to operate like global destinations, focusing on attracting guests not just to the hotel, but also to the country and city as a whole. Finding the right strategies to achieve that is essential.

Ed: I often think we're fortunate to be part of the hotel and hospitality sector because it's such an engaging product. Whether you're at the pub or standing in line with friends, one of the first topics of conversation is often about recent travel experiences: 'Have you been anywhere nice recently?' or 'Where did you go on holiday?' These discussions come naturally and our challenge is to be part of those conversations.

However, achieving this can be costly. We aim for a full-funnel marketing strategy to ensure that when someone mentions destinations like Dubai or New York, or even a lesser-known location in Germany, Marriott Bonvoy is the first name that comes to mind. This requires us to provide value throughout the entire customer journey. And, let us not forget, the core of our business still revolves around the stay itself. We must avoid over-promising and under-delivering at any point in the customer journey. Marketing needs to communicate its messages effectively, because the moment that communication falters we risk losing advocacy in those casual conversations with friends.

Travel and hospitality are topics that come up in every family room, restaurant or bar, but delivering consistently on our promises across the globe is crucial. I've spent nearly 30 years in this industry and I've loved every moment of it because I appreciate the way people talk about travel, hotels, stays, restaurants and experiences. If I were in a different industry, I might not have that same connection to the product I'm promoting.

I believe this passion is essential, especially for younger professionals entering the field. They should have an innate desire to travel and share their experiences, as this will significantly aid our strategic planning for campaigns. When they understand what the desired outcomes are, it alleviates some of the challenges associated with campaign planning. Conversely, if they lack affinity for the industry, it becomes much more difficult for them to grasp our goals for any given campaign.

Aligning global presence with local relevance

Stephen: It's hard to believe you've been in the industry for 30 years! Among all the campaigns you've worked on, which ones had a stronger youth element? Are there any that particularly stood out as ones you enjoyed the most, or perhaps ones that, while you didn't enjoy as much, still achieved significant success with Bonvoy, Marriott, or specific brands within the portfolio?

Ed: Naturally, there's a recency bias when discussing our recent collaboration with Taylor Swift. The initiatives we launched in conjunction with her truly felt like we were part of a live conversation that captured the world's attention. Whatever she's doing, she's doing it right, and we were fortunate to be included in that dialogue.

I appreciate it when hotels execute their campaigns effectively and can showcase the KPIs that align with successful outcomes. This gives me a sense of fulfilment, knowing that our efforts were worthwhile. For example, our UK team has done fantastic work focused on families, particularly through partnerships with Roald Dahl. These collaborations create wonderful customer journeys for children and their parents, where you can see the joy in the parents' eyes as their kids experience delightful moments – whether it's through an afternoon tea or a thoughtful turn-down gift. Knowing that all this began as a marketing idea and was seamlessly integrated into operations to foster loyalty is incredibly rewarding.

I can point to numerous successful initiatives during my time at Marriott International and Marriott Bonvoy, but I've also worked with other organizations throughout my career and they all strive to do similar great things. I find it easy to engage in this work because, as I said, we have a truly cool product. It's a joy to be in this industry, even though budgets often limit what we can accomplish. Choosing the right initiatives for our budget while ensuring the target audience is appropriate is critical.

From a youth perspective, some of the campaigns we've executed with the Moxy brand have been particularly impressive. For instance, we ran a successful British road trip campaign that resonated well. These domestic initiatives are most relevant to our current conversation, but if we can develop concepts in the UK that can be replicated in other markets, that is especially gratifying. Seeing the fruits of our hard work come to life is immensely rewarding.

The importance of customer lifetime value

Stephen: It's been really interesting to hear you discuss customer lifetime value. This is a significant aspect that many brands, including destinations, often overlook. For instance, they might invest heavily to attract a backpacker, but they lack insight into whether that traveller will return in the future.

Hearing how Marriott can track this is fascinating. One of the key themes in my book is the importance of investing in younger demographics – whether that's children on school trips or young adults aged 18–35 visiting for a weekend getaway. Once these individuals develop a love for your brand, the marketing efforts directed towards them later in life focus on upselling rather than discovery.

While technology may not fully support tracking for destinations, where it is available, the ability to monitor this engagement is incredibly valuable.

Ed: While I can't disclose exact percentages due to data privacy, we are aware of the uplift in bookings for second, third and fourth stays and this knowledge is incredibly rewarding for us. It's not solely about upselling; rather, it's about the fact that, when customers return, they want more. They might choose to upgrade to a higher room category or add ancillary experiences.

Once you engage a customer on this journey – whether they're younger travellers or more established ones – you know you're on the right path. It serves as validation for all the efforts we put into our strategies. I believe the concept of customer lifetime value represents the marketing holy grail we're all striving for. The earlier we can initiate that conversation or engagement, the better it will be for our overall strategy.

Conclusion

This chapter outlines the intricate and layered strategies Marriott employs to cultivate brand loyalty, especially among youth and emerging travellers. From the integration of mobile-first solutions to leveraging high-profile partnerships like Taylor Swift and expanding experiential offers through Bonvoy, Marriott's approach is a forward-thinking, multifaceted loyalty framework. These insights underscore a critical point: fostering lifetime customer value begins long before a traveller books a luxury stay or attends a corporate event. It starts in youth, through memorable encounters and consistent, authentic engagement that encourages guests to deepen their relationship with the brand over time. For any business looking to create lifelong connections with its customers, investing in early, meaningful interactions isn't just smart marketing; it's what is needed for sustainable growth.

KEY TAKEAWAYS

1 **Marriott's approach to youth travel:** Marriott's diverse portfolio of brands caters to various demographics, with certain brands like W, Aloft and Moxy particularly appealing to younger audiences. Ed emphasized the importance of building brand loyalty early, especially as many younger customers start engaging with Marriott through family vacations.

2 **Tracking customer journeys:** With tools like Marriott Bonvoy, the company can track the customer journey from their first interaction with the brand to how they progress over time. This helps Marriott better understand customer needs and tailor marketing strategies to engage them in meaningful ways.

3 **Bonvoy experiences:** Marriott's Bonvoy Experiences initiative offers a wide range of redeemable experiences, from sporting events to music concerts like Taylor Swift performances. This helps Marriott engage a broader demographic and go beyond just offering hotel stays.

4 **Marketing across generations:** Ed highlighted how the Marriott Bonvoy app has revolutionized Marriott's mobile-first strategy, engaging both younger travellers and older generations by making it easier to plan and book stays. There is also a lot of importance given to aligning marketing strategies to different age demographics, using partnerships with Manchester United and Formula 1 to target certain audiences.

5 **Social media and digital transformation:** Marriott's social media strategy is tailored to the strengths of each brand, with an emphasis on engaging with younger travellers through platforms like TikTok and Instagram. The goal is to match the right message to the right audience.

6 **Customer lifetime value:** A major focus for Marriott is customer lifetime value, where the goal is to keep customers loyal to the brand for decades, from their first solo trip to family vacations and beyond. This journey is vital for Marriott's overall strategy.

9

Destination marketing and management best practices

In this chapter I explore the complex challenges of destination marketing and delve into how youth travel could present a sustainable and innovative solution. To help discuss this topic, I'm joined by Manolis Psarros, the CEO and Chief Strategist of Toposophy, based in Athens. With an impressive reputation as a thought leader in placemaking and tourism strategy within the European Union, Manolis shares key insights on how destinations can craft responsible, forward-thinking marketing strategies that address the pitfalls of over-tourism while embracing the opportunities brought by younger, more conscientious travellers.

Manolis lends his expertise to demystify the process of crafting tourism strategies that incorporate youth travel while maintaining the economic and cultural vitality of a destination. Drawing on case studies, research and his extensive experience in tourism policy and placemaking, he provides a roadmap for how destinations can pivot their marketing approaches to attract visitors in a way that benefits both locals and travellers alike.

Understanding Toposophy: A place-based approach to tourism strategy

Stephen: Do you want to just explain what Toposophy does.

Manolis: The name 'Toposophy' itself encapsulates our philosophy. It derives from two Greek words: *Topos*, meaning place and *Sophia*, meaning wisdom. As a placemaking and marketing agency, we aspire to be recognized as leading experts in shaping places and empowering communities through strategy, research, marketing and events. Our approach considers the entire place, aiming to implement innovative strategies and research when collaborating with destination clients. Central to our work is the principle of prioritizing the needs and perspectives of local residents, ensuring that the strategies and solutions we develop are sustainable for the communities, as well as for the travellers who visit.

Stephen: The term 'placemaking' has evolved significantly over the past few decades. While some destinations are adopting an anti-tourism stance due to over-tourism, others, perhaps just a short distance away, remain relatively undiscovered. Youth travel constitutes a substantial segment of the tourism market. Do you believe that promoting youth travel could help alleviate some of the pressures from these major tourism hotspots? If so, have you encountered examples where destinations have focused on attracting younger travellers to distribute the visitor load more evenly?

Manolis: This has been a key topic of discussion in our industry. As someone dedicated to youth travel, you understand its significance beyond just a market segment. The crucial point is that we shouldn't view youth travel merely as another segment to analyse. Unlike other traveller demographics, youth travel can act as a powerful tool to balance the challenges of over-tourism across various locations.

It's essential to recognize that young people are often the pioneers, adventurers and explorers. They create lasting cultural connections with the places they visit, often in ways that transcend the typical tourist experience. In today's world, marked by geopolitical tensions and radicalization, fostering platforms that bring youth from diverse backgrounds together is crucial. This mission transcends the typical metrics associated with travel market segments.

Youth travel: A tool for balancing over-tourism

Stephen: When we define youth travel, it's important to consider both the age range and the purpose behind the travel. For instance, whether it's an organized school group visiting the Acropolis for the first time in Athens or a young couple enjoying a romantic week in Mykonos, destinations must recognize that younger travellers, regardless of their age, often have a desire to explore, experience and seek adventure. Ensuring accessibility and awareness of these needs is crucial for attracting this demographic.

Manolis: Indeed. There exists a significant perception gap among many of the destinations we've collaborated with. Surprisingly, national tourist boards and city destination management and marketing organizations (DMOs) may not fully recognize the various subsegments within youth travel. Many of them tend to focus on the stereotypical aspects, such as attracting young people for spring break fun or backpacking adventures, underestimating the potential of educational and cultural travel. Conversely, destinations in Western Europe, North America and Australia have a much clearer understanding of the educational aspect and have developed specific programmes, products and strategies to attract these travel segments. This highlights the considerable knowledge gap that needs to be addressed to allow some destinations to make themselves more accessible to young people.

QUESTION

Does your destination or brand actually have a youth-focused marketing strategy? If not, why not? Think about what your destination or brand has to offer and start creating a roadmap for how you need to build up your marketing to attract this valuable sector.

Stephen: You've worked with destinations across the globe, primarily in Europe. When advising on youth travel and various other segments, do you believe that destinations take a long-term approach? One of the central themes of this book is that investing in youth can lead to lifelong customers. Do you find that many destinations focus on quick wins, or do they recognize the potential for long-term gains? For instance, when they consider youth travel, do they understand that it represents a short-term cost with the prospect of significant long-term benefits over the lifetime of that traveller?

Manolis: There isn't a one-size-fits-all answer to this question. The approach really depends on the type of organization within the destination and the mandate they hold. When working with a national tourism board, the focus tends to be on destination marketing campaigns that yield short- to mid-term results. In contrast, when collaborating with local governments or a city DMO, the emphasis shifts to destination development, planning and management, allowing for longer-term strategies and solutions.

So, it's not necessarily about the destination itself; rather, it's about the organization you're engaging with and the specific objectives they aim to achieve.

Tourism ecosystems: The interconnected framework of a destination

Stephen: Are we referring to the 'tourism ecosystem' when we discuss this? This encompasses the entire framework of a tourism destination. For instance, if a destination aims to focus on youth travel but lacks the necessary visa policies to facilitate young people's travel, or if they want to attract school groups but lack adequate coach parking, those are critical considerations. Is that part of what you advise destinations on? Additionally, if they want to cater to the senior market (over 50s), do they have proper facilities for disabled access? It's all interconnected, isn't it?

Manolis: Absolutely – that's a trend we've observed in destination management globally over the last decade. More organizations and city agencies are integrating to provide comprehensive solutions that address management and planning issues while effectively promoting the destination to specific market segments. Previously, the structure of the industry and destination governance did not allow for such integrated strategies, making it challenging to address the needs of youth travel or other market segments.

Destinations like London, Amsterdam and Berlin are already leading the way in this direction, setting a trend for others to follow. For example, marketing organizations such as London and Partners, Berlin Partners and Amsterdam and Partners exemplify these integrated approaches. These organizations are working to respond to the challenges faced in tourism and are paving the way for more effective destination management.

TIP

Does your destination have an integrated approach to tourism? Looking into the ecosystem for the whole tourism sector in your region and then breaking it down into the different niches, will allow you to create more effective strategies.

Greece's potential in youth travel

Stephen: You mentioned various countries, but since you're based in Greece, do you see Greece – or its sub-destinations – as having a focus on youth travel? If so, what initiatives or features are being implemented? Conversely, if you believe there's a lack of focus, what do you think are the reasons behind that?

Manolis: I wish I could offer a more positive outlook on Greece's initiatives for the youth market. While there have been some recent efforts, particularly in marketing, from the national tourism board and a consortium of public and private agencies aiming to attract digital nomads and youth travellers more effectively than in the past, there is still much work to be done. In practical terms, Greece needs to enhance its connectivity with the digital nomad community and youth travel operators. To truly compete as a destination, the country must implement more substantial measures.

The impact of politics and policy on tourism

Stephen: Do you think this lack of focus is due to political issues, like how politics can impact visa regulations? For instance, in the UK, Brexit has had significant political implications, and recent news indicates that the UK is facing a tourism deficit of about £3 billion for the first time. Is it a matter of politics, or is it primarily a lack of funding tied to the Greek economy? Alternatively, as you mentioned earlier, do you think it's more about a general misunderstanding of the need for a dedicated youth travel strategy?

Manolis: You've hit the nail on the head. It's indeed a combination of various factors, but a significant part of the challenge stems from a limited understanding about how to approach youth travel effectively. If you have the knowledge of how the youth travel market operates, you'll be better equipped to navigate the political and other challenges that may arise.

The political aspect is crucial as well. Unfortunately, youth travel and digital nomadism have often been associated with issues related to affordable housing and short-term rental regulations. This association has led regulators to target these market groups more than necessary.

However, as we've discussed, the underlying issue is primarily one of knowledge. When you have access to comprehensive data, you can formulate balanced policies that avoid conflict.

> **TIP**
>
> If your destination has a healthy amount of youth travellers, but there seems to be a swell of bad press, do you have the stats and figures to explain the real situation? Educating government and local communities is essential in protecting a vibrant youth sector.

Building long-term affinities through youth travel

Stephen: I suppose I'm showing my age a bit, but you were once a student traveller, coming over to the UK for your bachelor's and master's programme. I've talked with you about how that experience has left you with a lifelong affinity for the UK. You are a prime example of how studying abroad can transform a person's relationship with a country, leading to lasting connections. Your journey has now evolved into reinvesting in the UK through your work, which is a testament to the value of those early experiences.

Manolis: This connects back to my earlier assertion that youth travel is more than just a market segment. It promotes cultural exchanges among young people globally, whether for educational purposes or other enriching experiences. This was certainly true in my case. Toposophy primarily serves clients in the UK, with 70–80 per cent of our clientele consisting of UK government authorities. Interestingly, I find it noteworthy that I often identify more with British values and principles than with Greek ones, which reflects the significant impact of my experiences in the UK. This perspective stems from my ability to study and travel affordably in the UK during my younger years.

Stephen: If the UK had implemented the policies, it has today back then, you might have considered going to Ireland to establish your business instead. The lifelong affinity that develops from having those special moments during significant years of your life is profound. The British government didn't need to advertise for you to return and set up a business; your experiences spoke for themselves.

Manolis: You're perfectly right on that. On the contrary, Brexit brought a lot more challenges than before for those of us who want to do business and operate in the UK compared to other countries. That said, both the UK and Ireland remain significant markets for us, where we've built strong working relationships and completed several important projects. Our presence in these countries highlights not only our affinity with their business environments but also the flexibility required to navigate a shifting political and economic landscape

---

TIP

In the long term, youth travel can create more economic benefit in the form of investment from abroad and even job creation.

Stephen: If a tourist board or destination is questioning the value of a TikTok campaign aimed at attracting digital nomads to a remote area, they may be overlooking a significant opportunity. Some might argue that these travellers don't contribute meaningfully, or that welcoming more international students could inflate housing prices. However, these young individuals could very well be the ones investing in and developing housing for the local community in 20 years' time.

Manolis: This highlights a broader issue regarding the appropriate key performance indicators (KPIs) for national and local organizations as they develop their strategies. Historically, these KPIs have often prioritized short-term metrics such as visitor numbers and quick economic gains. However, this approach tends to overlook the more profound, long-term benefits that travel and cultural exchange can provide, not only in financial terms but also in terms of social wellbeing and community resilience.

There is now a growing movement within the destination management sector to redefine success through a broader, more inclusive set of KPIs that also measure social impact and community wellbeing. As highlighted in our report, *Redefining Success: How DMOs can drive social and community wellbeing*, many DMOs are restructuring their frameworks to focus on sustainability, inclusivity and the wellbeing of local communities.[1] These new KPIs encompass a wide range of social benefits, including fostering stronger community ties, enhancing local quality of life and creating more inclusive tourism opportunities. However, much work remains to be done in fully integrating these into standard practice across the sector.

Challenges and opportunities in youth travel

Stephen: It's interesting to observe the varying emphasis on tourism across different countries. Over the past 10 to 20 years, tourism and politics have become increasingly intertwined. This growth has often been unstructured and entrepreneurial, leading to challenges that weren't foreseen or experienced two decades ago. With that in mind, what challenges do you currently see for youth travel, whether on a global or regional scale? Or do you believe that the opportunities for this sector are so significant that they can effectively overcome the challenges they face?

Manolis: Ah, you're the optimistic one! That's a question where you might already know the expected response. I believe the travel sector, including youth travel, is at a crossroads. Rising costs, visa hurdles, geopolitical tensions and environmental concerns pose significant challenges not just for youth travel, but for the entire industry. However, when we focus specifically on youth travel, I see the greatest opportunities emerging.

Young travellers are boundary-breakers, change-makers and pioneers eager to explore off the beaten path and discover new places and communities. This spirit creates a unique potential for growth within the sector. That said, destinations must be proactive. The youth travel market is highly dynamic, with constantly changing needs and expectations. To remain relevant, destinations must stay ahead of trends and ensure that their offerings align with what young travellers seek.

Stephen: As we discuss the evolution of destinations evaluating their tourism products, it's crucial to adopt an integrated approach that includes small businesses, large hotels, hostels, short-term rentals and accessibility – be it through easier visa processes or other means. The lines between demographics and age groups are increasingly blurring in terms of what travellers desire.

If destinations were to turn off the major tap of youth travellers, specifically those aged 11–35, it could create significant challenges in attracting older demographics, such as those in their 40s. Youth travellers often serve as the explorers, learning about a destination and forming lasting connections. However, for older travellers, if they've already experienced the destination through a school trip, educational programme or even a long weekend getaway, the dreaming phase has already been fulfilled. Your task then shifts to teaching them about the different facets and offerings within that destination, making their return visits more about deepening their understanding rather than discovery.

Manolis: In marketing, breaking down stereotypes and addressing perceptions about places and destinations is always challenging across various age groups and market segments. While it can be difficult to keep pace with the evolving youth travel market, it also presents a unique opportunity. This is essentially a blank canvas where we can craft the narrative of the destination we represent and invite young travellers to explore it in new and engaging ways.

Athens: An emerging hub for digital nomads

Stephen: I didn't travel to Greece growing up, but I have visited for work and I think the first time was when I was under 35. Since then, I've actively encouraged others to explore Athens. One of the challenges we discussed a decade ago was that many travellers, especially from the UK, tended to skip Athens in favour of the islands. However, I've noticed a positive shift lately, particularly with the influx of younger digital nomads who have chosen Athens as their landing spot in Greece. This trend has been beneficial for the Athens tourism sector.

Manolis: Indeed, and you might remember that the first time we met was during a workshop focused on Millennial travel and their behavioural patterns. Many of the topics we discussed back then have evolved into tangible trends in various destinations over the years. I can confirm that Athens is organically emerging as a prominent digital nomad destination. I emphasize the term 'organically' because, at this stage, it hasn't been part of a targeted strategy aimed at maximizing results. Instead, it's a second-tier destination that is gradually evolving into a first-tier one, now becoming very attractive to digital nomads compared to others that may not provide the same welcoming environment.

Stephen: As with any destination, access to workspace, reliable internet, convenient transportation and flexible living arrangements are crucial. However, when a destination's growth is more organic rather than structured, it can present challenges. Without the necessary infrastructure in place, frustrations can arise. Consequently, digital nomads, who often travel in groups or 'tribes', may seek out other locations that better meet their needs.

Manolis: In recent years, as with many major cities, Athens and other popular Greek destinations have often exceeded their carrying capacity during peak periods. However, this strain is largely attributed to other visitor types rather than digital nomads. Greece, particularly Athens, faces a distinct challenge when compared to Central and Western European cities, where travellers enjoy a wide array of transportation options. In contrast, Greece remains heavily reliant on air travel for the majority of international visitors, which has historically been a strategic disadvantage.

Nevertheless, once travellers are in Athens, the city effectively serves both as an end destination and also as a hub for exploring other islands or mainland destinations. Athens uniquely combines the quality of life and working environment of a big city with the ability to access a Greek island experience within one to two hours. This is a rare combination that makes it an appealing destination.

Stephen: It's quite unique. Many readers might not realize that Athens has an emerging digital nomad community, especially when the focus is often on the Acropolis and other historical sites. I was pleasantly surprised during my first visit for the youth travel summit. We participated in workshops and explored coworking spaces, uncovering aspects of the city that I hadn't anticipated. Interestingly, we hardly visited any of the historical sites during that trip!

Manolis: We primarily experience the urban lifestyle, much like youth travellers do. As you've noted, the accommodation infrastructure catering to these market segments has significantly increased over the years. There are dedicated service operators emerging to meet the needs of these groups, leading to organic development in this area. However, I believe there is still a lack of supportive platforms and specific tools that the government could have implemented for these segments, especially in comparison to other countries.

Stephen: That's fascinating! Considering your global perspective, what do you think will be the next major youth travel destination? Do you foresee any shifts in trends regarding how people are travelling?

Manolis: I believe that, after nearly 50 years of prosperity, wellbeing and peace, we will see many borders redefined in the coming decade. These changes will significantly impact the destinations that youth travellers and other travellers feel safe to visit.

Given the rising costs and geopolitical tensions, I think youth travel will increasingly focus on continental trips rather than long-haul journeys. While this trend used to be common, it remains to be fully defined.

Data-driven tourism strategies: A game changer

Stephen: As you mentioned, data is becoming increasingly important for destinations to effectively understand their markets. Instead of merely targeting luxury travellers because it's perceived as desirable, destinations should utilize data to tailor their marketing strategies to specific demographics. I believe this will become a significant trend as data accessibility improves for destinations worldwide.

Manolis: It's no longer just about marketing. In destination management, the mantra is clear: you can't manage what you can't measure. The focus now should be on effective destination management rather than solely on marketing efforts. Unfortunately, many government organizations tend to adopt a narrowminded approach, limiting their focus to traditional KPIs. They need to broaden their horizons and consider a wider range of KPIs that may not be directly related to tourism but can significantly impact their products and overall destination appeal.

Stephen: Data plays a crucial role for both operators and policymakers. It's essential to understand the long-term implications of tourism trends, especially when considering strategies for digital nomads. As you pointed out, the presence of digital nomads in Athens over the past decade has undoubtedly impacted the local infrastructure, presenting both challenges and opportunities. Recognizing these dynamics and developing a strategic approach is vital for sustainable growth.

Manolis: Indeed, to make informed decisions regarding infrastructure development or long-term strategies, we must broaden our data collection methods. Fortunately, AI is revolutionizing this process. We advise our destination clients to recognize that, while the findings may sometimes be challenging, the focus should be on analysing this data to develop effective solutions. It's not about promoting the findings or engaging in public debates that might lead to political conflicts or radical opinions; rather, it's about understanding the data and making the right decisions based on it.

Stephen: That's a perfect note to conclude on. Another recurring theme throughout these interviews is that data truly is paramount. The sooner destinations can access better and more real-time data, the more effectively they can strategize to meet the specific needs of their location.

Conclusion

The conversation highlights the importance of youth travel as a tool for sustainable tourism and cultural exchange. Destinations that fail to understand and adapt to the evolving needs of young travellers are missing significant long-term opportunities. Better data collection, political awareness and infrastructure development are crucial to making destinations more attractive and accessible to this demographic.

KEY TAKEAWAYS

1 **Youth travel as a solution:** Manolis discusses how youth travel could help alleviate over-tourism in popular destinations. He highlights that young travellers are often pioneers and explorers who establish deep connections with the places they visit. He stresses that youth travel should not be seen solely as a market segment but as a broader cultural exchange that can distribute tourist pressure more evenly across different areas.

2 **Challenges for destinations:** Many destinations, particularly tourist boards, may not fully appreciate the diversity within youth travel. They often focus on stereotypical ideas like party tourism while overlooking educational or cultural travel, which could be more beneficial in the long term. Manolis points out that destinations need better knowledge of the youth travel market to cater effectively to this demographic.

3 **Long-term vs short-term strategy:** There is often a mismatch between short-term and long-term goals in tourism strategies. While national tourism boards may focus on immediate gains through promotional campaigns, local DMOs often take a longer-term approach to destination management. The integration of long-term planning, infrastructure development and strategic tourism marketing is essential.

4 **Placemaking and infrastructure:** The concept of placemaking extends to ensuring that destinations are accessible, welcoming and equipped with the necessary infrastructure for different demographics. This includes visa regulations, coach parking and disability access, all of which need to be part of comprehensive tourism strategies.

5 **Greece and youth travel:** Manolis mentions that, although there have been recent marketing efforts to attract digital nomads and youth travellers to Greece, there is still significant room for improvement. Issues like connectivity, visa regulations and political challenges have hindered Greece from fully realizing its potential in youth travel.

6 **The importance of data:** One recurring theme in the discussion is the need for better data collection to inform tourism strategies. Destinations need to use data to measure the long-term impact of youth travel and adjust their key performance indicators accordingly. Understanding the long-term economic and cultural benefits of youth travel requires a shift in mindset.

7 **Geopolitical tensions and rising costs:** Manolis predicts that, due to geopolitical tensions, rising costs and environmental concerns, youth travel may become more focused on continental trips rather than long-haul journeys in the future. He stresses the importance of staying ahead of these trends and adjusting tourism strategies to meet changing travellers' expectations.

8 **Digital nomads and Athens:** Athens has become an organic hub for digital nomads, despite not having a specific strategy in place to attract this demographic. Manolis highlights the city's combination of urban life and proximity to islands as a unique advantage. However, he also notes that better infrastructure and support are needed to fully capitalize on this trend.

9 **Future of youth travel:** Manolis sees the future of youth travel as highly dynamic, with constant changes in travellers' needs and expectations. He believes that destinations that adapt to these trends – by investing in infrastructure, data and long-term strategies – will benefit significantly from the youth travel sector.

Note

1   Toposophy. Redefining success: How DMOs can drive social and community well-being, Toposophy, 2024. www.toposophy.com/insights/redefining-success (archived at https://perma.cc/8B29-M2B7)

10

Language travel

A catalyst for lifelong connections and economic growth

Language travel is far more than a fleeting holiday experience – it's a transformative journey that shapes individuals, industries and economies alike. In this chapter I chat with James Herbertson, co-founder of Bayswater Education, to explore the pivotal role this type of travel plays in developing global citizens. From teaching critical skills and fostering cultural exchange to driving economic growth, youth travel sits at the intersection of tourism and education. James shares his insights into how Bayswater has carved out a unique niche in this evolving market, balancing social impact with business success while offering life-changing opportunities for young people around the world. Through this conversation, we dive into the nuances of the sector: how it has adapted post-Covid, why work rights are a game-changer and why some countries are thriving while others are missing opportunities.

Stephen: Please can you give the readers an idea of what the company you founded, Bayswater Education, does.

James: Of course. Bayswater Education is a global upskilling business operating across three continents in five countries: Canada, the UK, France, Cyprus and South Africa. We provide a range of programmes, from language courses – like English in English-speaking countries and French in France – to professional development courses. These include digital marketing, luxury brand management and even specific skill sets like customer service for the hospitality sector, which we're launching in January.

Our programmes cater to people asking, 'What's next in my life?' We attract a diverse range of students. For instance, high school students might join us for one or two weeks, combining tourism with education – exploring a new country, learning about its culture and having fun. Gap year students, whether before or after university, come to us to enhance their confidence, communication skills and cultural awareness.

We emphasize soft skills, such as critical thinking, self-awareness and handling difficult conversations. These skills are invaluable for personal and professional growth. Our destinations are part of the appeal: iconic cities like Paris, London, Cape Town and Toronto. These are not just exciting places to visit but also impressive additions to a CV.

The dual role of youth travel: Bridging education and tourism

Stephen: It's fascinating how you have positioned Bayswater in the global education landscape. How did the journey begin?

James: My business partner Stephan and I both come from a language travel background. Seven years ago, we recognized that while language learning remained important, the focus was shifting. People increasingly wanted the experience of studying abroad without necessarily committing to a degree or master's programme.

In a post-Covid world, this desire has only grown stronger. People crave time in another country, meeting peers from around the globe and gaining valuable life and career experiences. We saw an opportunity to move beyond traditional language schools – which, frankly, the market didn't need more of – and focus on professional and experiential courses.

Stephen: Your focus on professional courses and global exposure is unique. How do these programmes prepare students for the future?

James: Our programmes are designed to equip students with skills they'll need both professionally and personally. Alongside traditional academic learning, we emphasize soft skills like communication, teamwork and critical thinking.

For example, a student might study digital marketing in London, gaining not only technical expertise but also practical experience collaborating with peers from different cultures. These experiences build confidence and adaptability – qualities employers value highly.

Additionally, our courses provide students with exposure to global destinations that resonate on a CV. For instance, studying in London or

Paris signals international experience and a broader perspective, which can set someone apart in competitive job markets.

Stephen: Bayswater has a strong social impact ethos, particularly with your one-for-one commitment. Can you elaborate on that?

James: When we founded Bayswater, we knew we wanted to make a difference. Quality education is a United Nations Sustainable Development Goal we feel passionately about. For every student who studies with us, we support another from a less privileged background.

Currently, we are running projects in the Middle East, Colombia, Brazil and Mexico, helping 12- to 16-year-olds learn English. These are often refugees or kids from disadvantaged areas. Our goal is to create a 'virtuous circle' rather than a 'vicious circle' where learning English opens doors to travel, tourism or international trade, empowering them to break cycles of poverty and access opportunities otherwise unavailable.

Stephen: That is very inspiring. How does Bayswater balance its social commitments with running a global business?

James: It's a core part of who we are, not just an add-on. The one-for-one model is woven into our business model. It's not just about financial sustainability but creating meaningful impact.

Our team and students embrace this mission and it is part of what attracts people to study with us. They know their experience isn't just benefitting them but also someone else who might not have had the same opportunity. It creates a sense of shared purpose, which is incredibly powerful.

TIP

Does your brand or destination have a strong environmental, social and governance strategy? The younger market are very conscious of social impact so maybe consider implementing one or marketing what you do offer more obviously.

From short courses to long-term journeys

Stephen: James, Bayswater Education is such a dynamic business and I think many people reading this might not realize how multifaceted companies in the youth education and travel sector really are. Often, when people see groups of kids with brightly coloured backpacks in cities during the summer, they assume it's just about language learning. But clearly, it's far more dynamic now.

Earlier in the book, I tried to define youth travel as covering ages 11–35. The lower end includes school trips for kids travelling without parents and the upper limit is tied to opportunities like Working Holiday visas. What's the typical age range of students studying with Bayswater?

James: That's a great question and I think your definition aligns well with what we see. However, we segment our audience into distinct groups because their needs and interests vary so much.

First, we have students aged 12–17, who make up a significant part of our market. These students typically come for one to two weeks, often during school holidays or as part of a school trip. For instance, if they're coming to London, their mornings might involve English language classes, while afternoons are packed with activities like visiting Buckingham Palace, the Science Museum, or taking day trips to places like Windsor, Oxford or Brighton.

Then there's a growing trend of secondary school students joining programmes that focus less on language and more on cultural or professional exposure. For example, we launched a global leadership programme where students can explore soft skills like teamwork, problem-solving and communication. These types of courses also serve as an introduction to studying abroad or even exploring a specific career path, such as medicine, law or emerging fields like AI.

Stephen: It's interesting how these programmes offer more than just language learning – they're setting students up for the next stage of their lives. What about the older demographic, say 18–35?

James: For the 18–35 age group, it's a mix. Some students are looking to bridge a gap between secondary school and university or between university and starting a career. Others might already be working but want to take a career break to upskill, build confidence or explore a different field.

This group thrives when they can stay longer in a destination, as living and studying abroad offers much more than a short leisure-based visit. However, the ability to combine studying with work rights is critical. In many countries, international students can work up to 20 hours per week while studying, which makes longer stays more feasible.

Unfortunately, we've seen how restrictive policies can negatively impact youth travel. For example, in the UK in around 2010, changes to work rights for language students significantly affected the market. Canada and Australia are now introducing caps and exploring similar restrictions, which could reduce opportunities in those markets. In contrast, countries that adopt policies like the UK's Youth Mobility Scheme – if it's expanded to key study abroad markets – could attract and support this demographic.

Stephen: It sounds like work rights are a game-changer for a destination looking to attract more young people looking to study on your sort of programmes. How do these opportunities benefit not just the students but the countries offering them?

James: Work rights make youth travel far more sustainable and beneficial for everyone involved. For students, it's not just about earning money – it's about gaining real-world experience, building confidence and learning new skills.

Take my own experience: I studied in Malaga, Spain and was lucky to work there as well. This was pre-Brexit and I ended up living in Spain for two years. It wasn't just about improving my Spanish; it was about personal growth and discovering a new culture. That experience ultimately shaped my career.

The reciprocal nature of these programmes is often overlooked. When young people from one country gain skills abroad, they often return home with new perspectives and abilities that benefit their communities. It's not just about attracting international students – it's about giving young people opportunities to explore, grow and bring those experiences back home.

Stephen: Your own story highlights the long-term benefits of international education. It's not just about the immediate experience – it's about how it shapes lives and careers. Do you see governments recognizing this value enough?

James: Not always, unfortunately. Many governments focus on immigration or short-term economic impacts, overlooking the broader benefits. However, forward-thinking countries with a bold vision for development understand that youth travel builds a globally skilled workforce.

When students return home, they bring back more than just language skills. They return with soft skills, industry knowledge and a better understanding of global markets. These are the people who drive innovation and economic growth in their home countries.

I think there's a real opportunity for countries to take a more holistic view. By supporting youth travel and education, they're not just creating opportunities for incoming students – they're enriching their own workforce and fostering stronger international ties.

Stephen: It's fascinating to see how youth travel has evolved into something so much bigger than just studying abroad. It's about building connections, skills and confidence that last a lifetime, as well as a lifelong connection with the destination where you studied.

James: Exactly. Youth travel today is about shaping global citizens. Whether it's through short cultural programmes, professional courses, or work–study opportunities, it's all about giving young people the tools they need to succeed in a connected world.

Planting the seed: The early impact of youth travel

Stephen: It is interesting, James, because, during an interview with Emma English of BETA, we touched on how youth travel often acts as an incubator for entrepreneurs. It's fascinating how those experiences – whether working, studying or living abroad – plant seeds for new ideas. You're a great example of this, and I guess I am too.

What strikes me, though, is how hard it can be to explain the long-term benefits of investing in young people to brands and destinations. That affinity we build during those experiences lasts a lifetime. For me, it was backpacking in Southeast Asia – those three months on local buses, eating street food, created a connection that still calls me back.

Do you find that some countries are better at understanding the long-term value of youth travel than others? In the UK, for instance, it often feels like the focus is on the short-term economic gains, with young travellers viewed as budget-conscious visitors who don't contribute much. Of course, we know that's not true, but what's your perspective?

James: That's a great question, and I think you've hit on something important, particularly when it comes to the UK. I'd say the UK suffers from a certain complacency, largely because it's a very developed market in youth travel. We've had a strong position historically, both in the number of people studying abroad and those coming here to study.

However, when it comes to marketing the long-term benefits of youth travel – what you called 'soft power' – there's a lack of strategic vision.

This short-sightedness overlooks how a single experience can build an enduring affinity.

I like to use a musical analogy to explain this. Think about your music tastes – they're often shaped during your teenage years, say 13–18. For me, it's the Britpop era: Oasis, Blur, Pulp. That music is embedded in me and I'll always have a connection to it. The same thing happens when someone studies abroad during those formative years. Those experiences become deeply personal, tied to their identity, and they carry that connection for the rest of their lives.

If you studied abroad in the UK at that age, you'll always hold those experiences, friendships and moments dear. That kind of relationship is invaluable – and we need to better quantify and communicate this when advocating for youth travel.

Stephen: That's such a great analogy – music and youth travel really do stick with you for life. But are there countries that seem to understand this better than the UK?

James: Yes! A great example is Ireland. They've done a fantastic job leveraging their position, particularly with their approach to work rights. They've been very savvy in recognizing the gaps created by Brexit and other policies in the UK.

The result? Their study abroad market, particularly in English language education, has boomed. In fact, they're already ahead of their 2019 numbers. Compare that to Canada, which is down about 25 per cent in the same period and the UK, which has seen a similar decline.

Ireland has seized the opportunity, while the UK remains potentially complacent and at risk of missing a great opportunity.

Connecting the dots: The disjointed ecosystem of youth travel

Stephen: What's the danger of this complacency for the UK?

James: The real danger is that it's all connected – it's an ecosystem, and if one part falters, the whole system could collapse. For example, in the UK there has been a massive investment in purpose-built student accommodation to house international students, and UK universities are heavily dependent on international students to fund their operations. If we don't see youth travel as the starting point – planting the seeds for longer stays, internships and degree programmes – this ecosystem is at risk.

Youth travel isn't just about wealthy students attending boarding schools. It's about creating accessible opportunities for a wide range of young people to experience the UK. Whether it's a week-long trip or a short-term internship, these experiences build connections that last a lifetime.

Without a long-term vision, we risk losing out to countries that recognize the broader benefits of youth travel.

Stephen: It's such a good point, and something that many of those interviewed in this book have agreed with. Youth travel isn't just about today – it's about creating lifelong connections, affinities and even economic benefits for the future. It sounds like the UK needs to shift its mindset from short-term gains to long-term investment.

James: Exactly. It's not just about short-term spending by young travellers – it's about creating ambassadors for the UK. People who studied or travelled here when they were young will come back later in life, often with their families, and they'll speak positively about the country in their own networks. That kind of soft power is invaluable, but we need to invest in it now to see the benefits later.

It's not so much short-sightedness – it's short-termism – and that's what needs to change.

TIP

Does your destination or brand have a long-term strategy for youth travel? If not, remember the benefits of this sector are not just short term. We can see from what James is saying, that one should look further than the initial spend of that young person and remember the soft power and return spending that young people could bring to your brand or destination.

Stephen: James, thinking about your analogy about music as a parallel to youth travel again, it has really resonated with me. For the record, though we're similar in age, I grew up loving Stevie Wonder – that was even my nickname at school!

What is fascinating is how those early experiences, whether through music or travel, stay with us. From a study perspective, I discussed this earlier in the book with Sally from Tourism Australia. The Australians are among the best at understanding the power of youth travel as a destination. Their Working Holiday visa process and campaigns are so consistent and they really integrate youth travel into the economy in a meaningful way.

What they cannot quantify, though, is the long-term impact. Once people return home, they do not necessarily need marketing to influence their memories – it is already embedded. For example, I have a colleague currently in Sydney, a city I lived in for eight months many years ago. She has been sharing photos and it's incredible how quickly the memories come flooding back: 'Oh, I remember that place and that one too!' It's moments like those that make me want to go back without any external prompt.

Hopefully, this book will inspire travel brands, hotels and destinations to recognize the value of creating these welcoming ecosystems for young people. Because, as you said earlier, you can have the most amazing universities with beautiful purpose-built student accommodation, but if the visa systems don't work, nobody comes. And it is often governments where this disconnect happens – different departments not talking to each other, creating barriers rather than opportunities. Sometimes this is intentional but sometimes these are the result of other decisions that create unintended consequences.

James: You've hit the nail on the head. The true benefits of youth travel, including the soft power it generates, can take a generation to fully manifest. You need time for these young travellers to grow into decision-makers, leaders in government, or influential figures in their sectors. One of our values at Bayswater is 'think in generations, not numbers' and this serves as much as a metaphor to respond to your question.

A fascinating aspect of our business at Bayswater is that we have grown through acquisitions, incorporating legacy colleges into our brand. What is remarkable is how often, even decades later, someone will walk through the door and say, 'I studied here in 1978.' These are often people now aged 40 to 65, sometimes with their children or even grandchildren in tow.

When you talk to them, you realize the profound impact their time studying abroad had on their lives. Many have gone on to lead family businesses, become specialists in medicine or law, or even start their own ventures. They might not talk about their current lives, but their stories about studying abroad are vivid and full of passion. That experience is still a defining part of who they are.

These former students become advocates for the experience of studying abroad. They carry that love and nostalgia with them and it shapes their outlook in ways you wouldn't necessarily predict. That's the challenge – when you look at a student today, you might not immediately see the potential. But one day, they could be the next finance minister or CEO.

TIP

Do you remarket your brand to young travellers? Remember, if you deliver a great experience, they are likely to return so make sure you are collecting young visitors' data and use it for future campaigns.

Stephen: That's such a powerful perspective. It reminds me of a conversation I had with someone who said wherever they went in the world, they had met people reminiscing about their time studying in the UK – whether it was Hastings, Manchester, or elsewhere. Those experiences stick and they carry such a strong emotional connection.

James: Exactly. I remember a story shared at an 'English UK' Parliamentary event where a minister reflected on how often they met international leaders or professionals who said, 'Oh, I did my studies in the UK.' It's an incredible endorsement of the long-term value of youth travel.

But governments often operate on five-year cycles, which makes it difficult to appreciate the generational benefits. The immediate economic impact is important, of course, but what's harder to quantify is the lifelong goodwill, the relationships and the networks that are built. Those are the things that truly endure and bring the long-lasting value that is so important for the economy.

Stephen: You are absolutely right. The challenge is making those in power recognize that youth travel isn't just a short-term economic driver – it's an investment in future connections, ambassadors and advocates for your destination.

It reminds me of a conversation I had with Ed from Marriott Hotels in Chapter 8. He spoke about their loyalty scheme and how pivotal childhood experiences can be. For instance, kids who stayed at a Marriott resort at 12 or 13 carry those memories into adulthood. When they start making their own travel decisions, those nostalgic, positive associations can guide their choices.

It's fascinating to consider how powerful these early experiences are and how they create brand loyalty that can last a lifetime. I'd love to see someone quantify that because it seems like such an untapped area of understanding.

On that note, you've been immersed in language education for a while now, both as an entrepreneur and as a student. How has the sector evolved, especially post-Covid? What are young people looking to learn these days? Have you noticed any standout trends?

The value of in-person experiences

James: There are two key questions here. The first is about the impact of Covid on online learning, and the second is how the language education market has evolved.

Let's start with the first: how did Covid shift the focus towards online learning? Covid had an immediate and profound impact, particularly on how education was delivered. It supercharged the emphasis on online learning out of necessity. With borders closed and travel restrictions in place, we had no choice but to pivot. Online platforms and virtual classrooms became the norm almost overnight.

At the same time, the demand for language learning in person didn't go away. It's important to remember that many young people still had this desire to travel and experience new cultures – they just couldn't do it during the height of the pandemic. So, while online learning served as a stopgap, it didn't replicate the full experience. The sector quickly realized this. We learnt that for most students, especially younger ones, it's not just about the curriculum; it's about the immersion, the environment and the connections they form.

Stephen: And what about the second question? How has the language travel market evolved post-Covid?

James: That's where things get interesting. The market has seen several shifts over the last decade, and Covid accelerated some of these trends. First, the overall level of English proficiency around the world has improved dramatically. Take schools, for example. In the past, a teacher might only have had an A2 or B1 level of English on the CEFR scale, which made it hard for students to progress beyond that. But today, many teachers are near-fluent. Combine this with the rise of bilingual schools, after-school language programmes and the push from parents and grandparents for kids to learn English and you have a significant boost in global language skills.

However, what hasn't changed is the thirst for travel and cultural experiences. Young people still want to study abroad. Travel is one of life's great adventures and that hasn't diminished. But the way students engage with these experiences has evolved. For instance, the trend now is for shorter stays – two to four weeks rather than six months to a year. This is partly because many students arrive with higher proficiency levels. Instead of needing long-term immersion to reach fluency, they're looking to 'top up' their skills or gain confidence.

Stephen: So shorter courses have become more common?

James: Definitely. Students are also looking for more tailored and unique experiences. It's no longer enough to just offer a standard English course. Many students are saying, 'I can do that in my own country.' They want something that's different, something they can't get at home. For instance, they do not want to sit in a classroom with 15 people from their own country. They want to be surrounded by an international mix of peers, whether from Sweden or Saudi Arabia. They want to learn about digital marketing, luxury brand management or other micro-credentials that give them something tangible to take away – beyond just language skills.

Another big trend is the increasing number of younger students travelling to study. In the UK, for example, there are now more students under the age of 18 studying abroad than over 18. This started a few years ago and has only grown. The UK is seen as a safe destination and it's accessible for European, Middle Eastern and Latin American markets. Countries like Malta, Ireland and Canada are also seeing similar patterns.

Stephen: What about online learning? Did the shift during Covid have a lasting impact on how people learn languages?

James: Online learning was a necessity during the pandemic, but it has limitations. Many of us in the sector faced challenges. Our revenue virtually disappeared overnight because our business is fundamentally about in-person experiences. If we were designing an online business, we wouldn't invest in expensive real estate in city centres. But, more importantly, what sets language travel apart is the immersive experience. You simply can't replicate that online.

We have also seen that completion rates for online courses are very low. It is easy to start an online course but just as easy to stop. When students travel to a destination, they are committed. They show up, they engage and they complete the programme. Apps like Duolingo are great, but they are not a substitute for the experience of being in a new country, surrounded by the language and culture.

Stephen: So, you see technology as more of an enhancement than a replacement?

James: That is correct. Technology can absolutely enhance what we do and AI offers exciting possibilities. But the heart of our sector is about people and experiences. It is like Maslow's hierarchy of needs.[1] At the top, you have self-actualization – the realization of one's potential. That's what these experiences are about. Sitting in a room with people from 10 different countries, discussing global issues, debating and learning from each other – that is irreplaceable and cannot be replicated in an online environment.

It reminds me of the movie *Good Will Hunting*. There's a scene where Robin Williams's character tells Matt Damon's character, that you could know everything about the Sistine Chapel, but you've never actually stood there and looked up at it. That's the essence of what we do. Online learning can only take you so far. It's the real-world experience that stays with you for life.

Stephen: It's interesting you mention the shift towards experiences over material possessions. That's a recurring theme in this book and the experiences of the interviewees when looking at youth travel. There is this great book I have referenced before, *Stuffocation* by James Wallman, which was written in 2013. He predicted that the upcoming generations would value experiences more than things and it feels like that's really come to fruition.[2]

In Sweden, they have this concept called *Lagom*, which roughly translates to 'just the right amount' – having as much as you need, no more, no less. It extends to how they approach travel, too, focusing on experiences rather than accumulating physical things. It feels particularly relevant when we talk about youth travel and education. For me personally, working and learning while travelling has been far more impactful than anything material. I'll always remember cooking gumbo and jambalaya in New Orleans or making spring rolls with a grandmother while staying in a guesthouse in Vietnam. Those experiences stay with you.

When we think about youth travel and language learning, how big is this industry? Can you give us a sense of the scale, globally or in the UK?

James: While I do not have the global figures to hand, I can tell you that in the UK, the youth travel sector is often quoted by BETA as contributing around £27 billion to the economy. That's a massive number, but it's important to break it down to understand the full impact.

When people talk about this figure, they often focus on the primary beneficiaries, like the educational institutions where students enrol. But the real value lies in the *multiplier effect*. Think about the ripple effects: students need accommodation, and in many cities they end up spending more on housing than on the actual course fees. Then there's the money spent on entertainment, hospitality and transportation.

For example, you have the taxi drivers picking students up from the airport, the homestay families earning additional income from their spare rooms and local businesses benefitting from increased foot traffic. In some cases, a single homestay arrangement can help support an entire family in that local community. Then, add on the spending on local attractions – whether it's visiting museums, booking excursions or just exploring. These things all add up.

Stephen: That is true, and it feels like language travellers stay longer, which must amplify that impact.

James: Exactly. Language travellers tend to stay for weeks, sometimes months, rather than just a few days like business travellers or those on luxury holidays. That extended stay means their economic contribution is much greater.

Contrast that with, say, a luxury traveller attending a conference. Sure, their daily spend might be higher, but they're typically in and out of that destination in a couple of days. They don't have the same opportunity to explore the city or engage with the local culture. Youth travellers, on the other hand, immerse themselves in the destination.

Take Cape Town, where one of our schools is located, as an example. A student studying at our school there will almost certainly go on a safari, whether it is a day trip to a wildlife camp or a multi-day excursion to Kruger National Park. They'll visit the Western Cape, hike Table Mountain and tick off other bucket-list experiences. All of this creates an ecosystem of economic activity and benefit that stretches well beyond their tuition fees.

Stephen: It's an entirely different scale of engagement. It's not just about the educational institutions; it's about creating an entire economy around their stay.

James: That is why it is so frustrating when governments or tourist boards focus solely on luxury travel or conferences. Yes, those markets are important, but youth travel has a far-reaching impact that often goes unrecognized. It's not just about the students themselves; it's about everyone they interact with during their stay and the long-term connections they build with the destination. The challenge is quantifying this impact and getting governments and industries to truly value it. But if you think about how much youth travellers contribute – economically, culturally and socially – it's clear that they're a vital part of the tourism ecosystem.

Given that young people account for a third of all tourists to the UK you would like to think there would be more emphasis placed on the sector rather than high end hotels but in reality, this just does not happen.

Stephen: I suppose you are hinting that the UK is not a supportive environment to operators in your space compared to other countries. Given that you operate in different regions, are some of those places more supportive, whether it is a longer or shorter term approach to that support?

James: Certainly, and it's an interesting comparison because each country approaches youth travel differently, reflecting their broader attitudes towards tourism and education. For example, Canada has been very proactive in supporting youth travel and education. At the ICEF conference in Berlin, I attended an event hosted by the Canadian Embassy where they brought in representatives from Languages Canada and the ambassador himself to engage with operators. That kind of high-profile support makes a real difference in showcasing their commitment to the sector.

Similarly, South Africa has also shown strong support. Recently, a group of South African schools collaborated to create a dedicated space at ICEF Berlin, with backing from the South African Tourism Board. That kind of partnership between education providers and the national tourism authority helps build a unified message about the importance of youth travel to the country's economy and its global reputation.

The UK, on the other hand, can feel less consistent. There are pockets of support – Bournemouth, for instance, is an excellent example where the language travel sector is a core part of the local economy. Homestay programmes are well-known and widely embraced by the community and there is a stronger understanding of the sector's value among local leaders and politicians. However, this level of recognition does not always translate at a national level, which can leave the industry feeling fragmented.

Stephen: That's a really striking contrast. So, in places like Bournemouth, where the local economy is closely tied to youth travel, you see more engagement. But on a national scale, it seems like the UK lacks that same focus. Is that fair to say?

James: The UK does have its successes, but they tend to be localized rather than part of a cohesive national strategy. During Covid, for example, the language travel sector struggled to engage with local governments for support that was on offer to other businesses. In many constituencies across the UK, there simply were not enough schools to make our industry's impact visible compared to other sectors. As a result, we lacked the collective voice we needed.

Other countries, like Canada and Australia, seem to have a more unified approach. They understand that youth travel is not just about tourism – it's an investment in their country's future. The long-term benefits of attracting young travellers, whether they come back as students, professionals or tourists later in life, are enormous. Unfortunately, the UK often seems to focus more on traditional tourism or luxury travel, which overlooks the broader economic and social impact of youth travel.

Stephen: That's eye-opening. It sounds like countries that take a more collaborative approach – bringing together tourism boards, education providers and governments – are better positioned to support and grow the language travel market. Do you think that kind of unified strategy is what's missing in the UK?

James: Collaboration is the key. The countries that are thriving in this space recognize that youth travel spans multiple sectors – education, tourism and even cultural exchange. They actively invest in bringing those sectors together to create a cohesive strategy. That is something the UK could learn from. If we want to compete and attract young people who will contribute to our economy for decades to come, we need to move away from a fragmented approach and start thinking about youth travel as a critical part of our national strategy.

Stephen: In Chapter 4 with the travel blogger Kash Bhattacharya, we discussed how sometimes really small destinations have amazingly supportive tourist boards that really understand the power of tourism and what it brings to the local community, whether it is generating jobs or revenue, and that tourism is a general force for good.

It does feel like the UK and other countries around the world do not quite appreciate the total positive impact tourism brings generally and then in turn youth tourism more specifically.

We have obviously worked in the sector for a long time and what is actually coming to mind as we're talking is that maybe that's where the downfall is sometimes for youth travel – the issue is not the specific niche we operate in, but more broadly the country does not appreciate or quantify tourism as an industry.

As we mentioned earlier, for Bournemouth, which has a population of nearly 200,000 people, tourism is the second biggest industry. As a city, they do not want to lose that revenue generation, but maybe others not involved in the city do not realize the scale. If we look at Homestays we understand what it is, as we have worked in the sector, but maybe those not in the sector do not understand that importance or financial impact of this sort of accommodation.

They probably don't understand that in the summer in Bournemouth there's thousands of kids staying in homestays that are generating millions of pounds, but because the public do not see it, or cannot quantify it, it does not become an important thing from a local government perspective and therefore the destination is sometimes hamstrung by the national government.

The multiplier effect of youth travel

James: The missed opportunity, I believe, lies in not fully appreciating the interconnected ecosystem of youth travel products. If you take a moment to think about it, whether you're looking at education or tourism, our sector straddles both worlds. That duality is part of its challenge – we're not neatly defined by one sector and that often leads to fragmented strategies and missed connections.

Youth travel is about planting seeds early in the journey of someone's life. It's about creating an affinity for a destination that lasts well beyond the initial experience. For example, when we think of the hundreds of thousands of students who come to the UK every year to study at universities, how many of them started their journey to that decision years before? Perhaps they visited the UK for a two-week English language course in their teens. Maybe they prepared for their IELTS exam here or took a short cultural immersion trip. Others might have studied for a degree in their home country but spent a semester abroad or completed an internship in a foreign city.

These are not isolated experiences – they're part of a continuum. And we are at the very start of that pipeline, providing those initial experiences that build confidence and spark curiosity. Yet, there is a disjointedness in the way this progression is often viewed. Decision-makers fail to connect the dots: the student who studies abroad and returns to their hometown as an advocate, sharing their stories with 10, 20, or even 30 other people, has a multiplier effect. They inspire others – friends, family, colleagues – to seek similar experiences.

This ripple effect is what we risk losing if we don't nurture the youth travel sector as a foundational part of the broader tourism and education ecosystems. Cutting off those initial opportunities to explore, learn and grow diminishes the long-term impact that youth travel has on global relationships, personal development and economic growth. It's not just about today – it's about the future, and the missed opportunities are far too significant to ignore.

Conclusion

As James explains, youth travel is a sector with unparalleled potential – not just as an economic driver but as a force for good in shaping the next generation of global citizens. Beyond the immediate benefits of education and tourism, it creates lasting connections, cultivates soft power and opens doors for young people to realize their full potential.

While some nations, like Ireland, have embraced progressive policies that support youth travel, others risk stagnation by focusing on short-term gains. The UK faces challenges due to a lack of cohesive strategy, missing the opportunity to fully harness the multiplier effect of youth travel and language travel in particular.

KEY TAKEAWAYS

1   **Youth travel as a transformative experience:** Youth travel shapes individuals in profound ways, enhancing confidence, critical thinking and adaptability while fostering cultural awareness and global perspectives.

2   **The multiplier effect:** The ripple effects of youth travel go far beyond the individual traveller, impacting families, communities and local economies. These experiences inspire others to explore and build lifelong connections to destinations.

3   **The role of work rights in attracting youth travellers:** Progressive work–study policies, like those in Ireland, make destinations more attractive to young travellers. These opportunities not only support students financially but also provide invaluable real-world experience.

4   **Post-Covid shifts in learning and travel:** While online learning grew during the pandemic, it cannot replicate the immersive and transformative power of in-person experiences. Short-term, skills-based programmes have become increasingly popular.

5   **The need for strategic collaboration:** Successful countries and destinations recognize youth travel as a vital part of their education and tourism ecosystems, fostering collaboration between governments, education providers and tourism boards.

6   **A call for long-term thinking:** Governments and industries must shift from short-term economic thinking to viewing youth travel as an investment in future leaders, innovators and ambassadors. Without this, the sector risks stagnation and losing its competitive edge globally.

Notes

1   P Mcleod. Maslow's hierarchy of needs, Simply Psychology, 2024. www.simplypsychology.org/maslow.html (archived at https://perma.cc/8GWZ-BEL4)

2   J Wallman (2014) *Stuffocation: Living more with less*, Penguin, New York. http://static.booktopia.com.au/pdf/9780241183809-1.pdf (archived at https://perma.cc/T33A-EYVS)

11

The lifelong value of youth mobility

Youth mobility is a dynamic and transformative force within the travel and tourism industry, touching the lives of countless individuals and shaping global communities in profound ways. In this chapter, we dive into the expertise and insights of Vicki Cunningham, CEO of BUNAC, JENZA and USIT, who brings decades of experience in facilitating cultural exchange programmes. Through her work, Vicki has played a pivotal role in enabling thousands of young people to explore the world, immerse themselves in new cultures and acquire invaluable life skills.

Stephen: Obviously, as someone who worked abroad when I was younger, especially when I lived in Australia, what BUNAC does is something very close to my heart. I wonder if you could explain what BUNAC does as an organization and the other brands that are part of it.

Vicki: Absolutely, Stephen. BUNAC is the brand I first joined back in 2008, but its history stretches much further back. It was formed in 1962 by a group of university students, essentially as a university society, with the mission of encouraging reciprocal travel between the UK and US. The aim was to promote cultural exchange by allowing young people to immerse themselves in different cultures and gain first-hand experience of another way of life.

This idea of cultural exchange and creating reciprocal programmes has remained at the heart of BUNAC for over 60 years. The focus has always been on facilitating young people travelling into the UK and Brits going abroad to places all over the world.

Today, our work revolves almost exclusively around work-and-travel opportunities. All our programmes are centred on internships, summer camp roles or seasonal work. The goal behind these experiences is to provide an affordable way for young people to travel, see the world and engage deeply with local communities. These opportunities allow them to earn some money to support their travels while they're exploring and learning about a different culture.

During Covid, a turbulent time for our industry, BUNAC and USIT merged into one organization, resulting in an amazing opportunity for alignment. USIT is Ireland's leading work and travel brand, with a strong focus on US exchange. Both brands share a similar history with USIT, also set up over 60 years ago by student founders.

In 2022 the group launched a new brand, JENZA. JENZA took all the best bits of BUNAC and USIT – our people, passion and expertise – and shaped it for a generation of switched on and socially conscious travellers. A generation that didn't want to live someone else's adventures, but shape their own.

JENZA now solely sells all our working holidays and internships abroad globally, while BUNAC is focused on camp and USIT continues to fly the flag for work and travel in Ireland.

We cater exclusively to 18- to 35-year-olds, with our primary target demographic being 18- to 25-year-olds. Over the years, we've experimented with different destinations and offerings, but our core mission has remained constant: creating opportunities for cultural exchange and life-enriching travel experiences.

What's been particularly rewarding is seeing how this model has stood the test of time. Despite being around for over six decades, demand for our programmes has never been higher, proving that the desire to explore the world and connect with different cultures is as strong as ever.

The skills and lessons of travel

Stephen: It's really interesting to hear about the foundations of BUNAC. One of the themes Emma English from BETA mentioned was that youth travel often serves as an incubator for entrepreneurship. It's amazing how many people I've interviewed, and even how many people you and I probably know in this sector, who started out by participating in some sort of youth travel experience. Do you find there have been stories of BUNAC customers who went on to create their own businesses because of their experiences? Or is that not necessarily something you've tracked?

Vicki: Gosh, that's such a fascinating point, Stephen! Off the top of my head, I can't think of an exact example of someone who went on to become an entrepreneur directly because of their BUNAC experience, but I'd say the qualities they develop through these programmes certainly lay the groundwork for entrepreneurial success. Almost every participant we interview, survey or hear from anecdotally speaks about how these experiences have shaped them.

For example, the resilience they build, the confidence they gain and the open-mindedness they develop are recurring themes. Taking the initiative to step out of your comfort zone and navigating unfamiliar environments are such formative experiences – and they're fundamental skills for anyone thinking about starting their own business.

On top of that, these programmes teach participants how to work collaboratively with others, which is another cornerstone of leadership and entrepreneurship. Even if they don't end up launching a business, many of them go on to careers in team management, teaching or other roles that involve mentoring and supporting others. Youth travel really nurtures people skills – things like empathy, adaptability and cultural sensitivity – that are so valuable in both professional and personal life.

Honestly, the impact of these experiences is profound, and I think if we tracked it more formally we'd likely uncover some incredible stories of BUNAC alumni who went on to become innovators and leaders. My own team is a great example – all of them are alumni of these types of programmes, and every day I see how those experiences continue to shape their approach to work and life.

If I come across a specific case study, I'll definitely share it with you. But even without one at hand, I'd say the evidence is clear: these kinds of experiences equip young people with a toolkit of skills that make entrepreneurship – or any career ambition – a real possibility.

Stephen: That's such an important point, Vicki. One of our former interns, for example, worked with me in a hotel years ago – not exactly an obvious path – but now he runs the largest voucher platform in Benelux with 350 staff. Interestingly, he picked up the foundational skills and the business concept from his time in the UK, where he saw how Groupon was booming. It's a perfect example of how sometimes the hard, practical skills come from the work experience itself – whether it's being a personal trainer in a gym in Australia or waiting tables at a beach resort. But, just as often, it's the soft skills you develop that really make the difference. Finding an apartment in a foreign city, dealing with local tax laws, setting up a bank account or navigating cultural differences – all of those experiences shape you in ways you don't fully realize at the time.

And that's been such a common thread throughout this book: the idea that you truly learn by doing. And the 'doing' is the travelling – it's living in the unfamiliar, solving problems and pushing yourself beyond your comfort zone. That's where the real growth happens.

Vicki: Absolutely. You've hit the nail on the head. It's always hard to say this without sounding patronizing as someone in their 40s, but the truth is that it's often the moments of adversity, challenge and discomfort that leave the deepest and most valuable lessons. No one sets out hoping to miss an internal flight, lose their bank card in a foreign city or misplace their passport, but these things happen. And while they're incredibly stressful in the moment, they're not insurmountable.

The beauty of programmes like ours is that, while we're there to lend a hand when needed, more often than not it's the young person who has to step up and resolve the situation. That act of leaning into a challenge, of figuring things out on your own, builds an enormous sense of resilience. It's that resilience – the knowledge that 'I can solve things for myself' – that sets the stage for taking bigger risks later in life, whether in business, career or personal endeavours.

This isn't just about surviving mishaps; it's about the confidence that comes from thriving in unfamiliar situations. That confidence can make a young person more willing to take the road less travelled, to push themselves beyond the safe, easy options and to embrace opportunities that feel a little daunting.

And, of course, there's the exposure to new ideas and perspectives. Travelling means rubbing shoulders with people from all walks of life, and that alone is transformative. It opens your eyes to possibilities and ideas you might never have considered otherwise. Your colleague with the voucher business is a perfect example. His experience in the UK sparked an idea that he took back and adapted to his own context. That's the power of exposure – of stepping outside your bubble and seeing what's out there.

Ultimately, these experiences equip young people with a mindset of initiative and possibility. Whether they become entrepreneurs, leaders, or simply more empathetic individuals, the ripple effects are profound.

Stephen: Yeah, that's really exciting. You guys operate on a global scale and youth mobility often relies on reciprocal agreements between governments to make it possible. I'm curious – what destinations are you seeing as the most popular right now and what do you think is driving that interest?

The role of policy and programmes

Vicki: It's such an interesting question and there's definitely a lot to unpack. One thing to note is that youth mobility trends are often shaped as much by government policies as they are by traveller preferences. Since the programmes we offer are confined by various youth mobility restrictions, the destinations we see as popular may not always reflect the full picture of what's trending globally.

That said, the US continues to hold very steady as one of the top choices. Despite the complexities of its visa system and the political environment – we're just 20 days post-election as we speak – the country consistently appeals to young people looking for short-term working opportunities. There's something uniquely alluring about the US for these types of experiences, whether it's the culture, the perceived opportunities, or just the idea of living the 'American Dream', even if only for a few months.

We're also seeing growing interest in places like Japan and South Korea. These destinations are emerging as favourites for young people who want to push themselves a bit more, who are looking for a challenge or who are drawn to a cultural experience that's vastly different from their own. What's exciting is that these countries are actively investing in making themselves attractive to youth travellers. They're implementing programmes and incentives designed to lower the barriers to entry while still offering a sense of adventure and novelty.

Australia, of course, remains the perennial favourite, especially for young people from the UK. From a UK perspective, we're the number one market for sending young travellers to Australia. But, even on a global scale, Australia's popularity is immense. That's largely due to its extensive network of reciprocal agreements with countries worldwide. Australia has created an incredibly youth-friendly infrastructure with its Working Holiday visas, its openness to international workers and its thriving backpacker culture.

Ultimately, it's often the visa agreements and youth mobility parameters that dictate the destinations young people choose. Countries that actively invest in these types of programmes – whether through flexible visa policies, promotional efforts or structured cultural exchanges – are the ones that see the highest demand. In a way, the destinations that make the most effort to welcome and accommodate youth travellers are the ones that succeed in attracting them. It's a clear reminder of how important policy and investment are in shaping global travel trends.

Stephen: Yeah, we've covered this in several interviews for the book. Some destinations seem to understand the value of these programmes and invest in them wholeheartedly, while others seem to almost operate in the shadows – offering working holiday or student visa schemes but failing to promote them effectively. It's like they don't want anyone to know they exist. Naturally, that leads to lower demand and missed opportunities.

Vicki: If you look at Australia, for instance, they're the poster child for doing this well. They not only invest in their programmes but also actively market them. Australia truly understands the economic and cultural value of youth mobility and that knowledge drives their commitment to making these programmes successful. It's not just about issuing visas; it's about creating a holistic framework that attracts young travellers and supports their experience once they're there.

On the other hand, take the US, which I think is a fascinating case study. Programmes like the BridgeUSA initiative fall under a broader umbrella of exchange programmes, all of which have been around since the Fulbright-Hays Act of 1961. I'm sure you're familiar with it. What's striking about the US approach is their understanding of the *public diplomacy* value of these initiatives. They don't just see these programmes as a way to fill short-term labour gaps; they see them as tools for building long-term international relationships and cultural exchange.

The scale of participation is staggering – somewhere around 200,000 people annually take part in BridgeUSA programmes alone. And that's just one slice of the pie. These programmes exist within an even broader category of exchange initiatives, many of which are highly strategic. I was struck by a stat that I heard recently from Deputy Assistant Secretary Pasini at the US Embassy in London. She mentioned that, currently, one in three global leaders is an alumnus of a US exchange programme.

That really brought it home for me. It's not just about the immediate economic impact; it's about influence, diplomacy and fostering global connections. The US understands that these programmes don't just benefit participants; they benefit the country's standing on the world stage. For countries that fail to grasp the value – whether it's economic, diplomatic or cultural – they're unlikely to put much weight behind their programmes. And when there's no investment or infrastructure, you naturally see low demand.

Ultimately, it comes down to a country's priorities and their ability to see beyond the short term. Programmes like these are long-term investments in relationships, soft power and economic growth. Destinations that understand this – and act on it – are the ones that thrive.

QUESTION

Does your destination have a youth mobility scheme in place? If it does, is it promoted alongside the rest of the tourism industry as a key driver to increase visitor numbers ?

Stephen: Yeah, it's really interesting to hear. I think it's something that's come up in a few interviews – the sheer number of prime ministers, presidents or other global leaders who have studied or worked abroad before they rose to prominence. It's such a telling statistic. I mean, a third of global leaders is massive when you really think about it.

Vicki: It's incredible, isn't it? And when you consider the scale of influence that figure represents, it's almost hard to believe. But when you look at the infrastructure and thoughtfulness behind programmes like the US's exchange initiatives, it starts to make sense. It's no accident – it's by design. These programmes aren't just about individual enrichment; they're about shaping the future global landscape. For the US it's clearly a strategic priority, and it shows in the long-term impact they've achieved.

Stephen: One of the recurring themes in my conversations has been about how well destinations like Australia execute and promote these programmes. When I spoke with Sally Cope in Chapter 5, who was formerly at Tourism Australia, she mentioned how proud they are of their success in this area. But what's interesting – and no destination I've spoken to seems to have mastered this – is the ability to track the lifetime journey of that young traveller. Do they come back later in life to start a business? Do they invest in the country or promote it to others? It's almost like destinations could create an alumni network, not unlike a university alumni system, but for those who've lived, worked or studied abroad in their country. It could be transformative if destinations started tracking that lifetime value.

Vicki: That would be fascinating. Not just tracking whether they return as high-net-worth individuals to set up businesses or invest in the local economy, but also understanding how many visitors they bring back with them over the years. Do they return with friends a few years later? Do they bring their families back 10 or 20 years down the line? The ripple effect of that one person's experience could be huge.

And that's just looking at the economic value. There's also the soft power element to consider. People who work or study abroad rarely come back and say, 'Oh, it was okay.' It's usually, 'That was the best experience of my life!' They become advocates, telling everyone they know, 'You have to visit Sydney, Toronto, or San Francisco. I had an incredible time.' The countries that really excel in this space are the ones that fully understand not just the immediate impact but the long-term benefits of creating these life-changing experiences.

Economic and cultural impact

Stephen: I think a lot of companies and destinations overlook the long-term value of youth travel because they're fixated on the short-term economic returns. They see young travellers as budget-conscious, assuming their impact is minimal because they're not spending on luxury hotels or fine dining. But, as this book has explored, the reality is quite the opposite. Youth travellers tend to stay longer, explore more diverse areas and have a broader economic impact. Beyond that, if we could quantify something like 'two-thirds of youth travellers return at least twice before they turn 50' or highlight the leadership statistic that 'one in three become global leaders', it would completely reframe the conversation. If a country were run like a business, it would be a no-brainer to invest in this demographic because they offer both immediate and compounding value.

Vicki: Exactly. It's such a strange and short-sighted approach when you compare it to other industries. Imagine if the banking industry only judged young customers based on their value when they were 22 years old. That's ridiculous. Banks prioritize acquiring young customers because they know that, in a few years, these individuals will be taking out mortgages, applying for car loans and making significant financial decisions. The same goes for many service-based industries – they're focused on capturing the youth market because they understand that these young people are the high-value customers of the future.

Youth travel isn't any different. These young travellers will grow up to become business travellers, family vacationers and luxury tourists. And if a brand or destination makes a great impression on them when they're young, it's far more likely they'll return, either out of nostalgia or loyalty. That's what industries like banking have figured out – invest in youth now for long-term returns.

Beyond that, young people are trendsetters. They influence others around them – friends, family, even older generations. I've experienced this personally. After I travelled through Malaysia and a few other countries in Southeast Asia, I couldn't stop talking about it. My parents, who had never considered visiting that part of the world, were intrigued. I ended up taking them to Malaysia and they loved it so much they returned with their friends. It's that ripple effect. One young person can inspire a chain reaction of visits and, suddenly, a destination benefits from multiple layers of travellers.

The idea that youth travel isn't worth investing in is baffling, especially when no other consumer-focused sector takes such a short-term view. Most industries know that the youth market is the foundation for future growth. It's not just about what young people spend now; it's about who they'll become and the choices they'll make in the future. And destinations or brands that don't recognize that are missing a huge opportunity.

Stephen: Exactly, and I think that's part of the challenge and the opportunity. It's one thing to have the infrastructure for tracking someone through their lifetime as a customer, like with airlines or banks, where there's an inherent continuity in the relationship. But, when it comes to destinations, the challenge is much more complex. There's no centralized system that connects the dots between someone's first visit on a student visa and their subsequent visits over the years, whether for leisure, work or family.

You mentioned airlines, and I think they've really nailed it with the loyalty programmes connected to round-the-world tickets. You start collecting points on a OneWorld or Star Alliance membership and, before you know it, you're loyal to that brand because it's rewarding your behaviours. If destinations could find a way to emulate that kind of model – perhaps by integrating tourism boards with immigration systems like ESTA or ETA – it would completely transform how we measure the long-term value of youth travellers. It's not just about the initial spend; it's about the cumulative economic impact over decades.

QUESTION

Have you looked into the return customers to your brand and destination, and what age they started the journey with you?

Challenges and opportunities

Vicki: Trackability is key, but it also comes down to taking a long-term view. It's interesting when you look at products like mobile phones because people often do not change their number or provider and I'm exactly the same. I've been with the same provider since university. They came to campus, offered me a great deal, and I've never bothered to switch. That kind of loyalty isn't just about the product; it's about building a relationship from the start. The same applies to destinations. If you make a great first impression and create a seamless, positive experience, you lay the foundation for lifelong loyalty.

But that requires data, and with data comes proof. Once you have the numbers to demonstrate the long-term value of youth travellers, it becomes much easier to make the case for investing in them. Right now, a lot of destinations still focus on short-term economic metrics – the immediate ROI of youth travellers, which, as we've discussed, often doesn't reflect their true value. The challenge is getting countries and tourism boards to think beyond political cycles and short-term budgets.

Take the US as an example. Their exchange programmes have been running since 1961, thanks to the Fulbright-Hays Act. That's a programme with decades of legacy and infrastructure behind it and it's produced tangible results – not just economically but in terms of soft power. That kind of roadmap takes time to build, but once it's in place it's incredibly robust.

Australia's Working Holiday visa programme is another great example. It's been around for decades and they've had time to refine and expand it. The programme is hugely successful because it's backed by long-term vision and consistent investment. Countries and destinations that can think in those terms, rather than being tied to a four- or five-year political cycle, are the ones that will ultimately benefit the most from youth travel.

Stephen: I suppose, as business leaders, we're always trying to balance the short-term demands with the long-term vision. It feels like some destinations and brands haven't quite mastered that cycle. When you think about the impact of Covid, it wasn't just a disruption; it fundamentally halted youth mobility. People were stuck in countries – Australia was a big example of that – and for those who had dreams of travelling, like my wife and I with Japan, it was like having the rug pulled out from under you. It hasn't been easy to make up for that lost time since.

From your perspective, have you seen a shift in what drives people to work and travel since Covid, or has the motivation remained the same?

Vicki: I think, if anything, the demand for travel has rebounded stronger than ever. When something is taken away, you suddenly realize how valuable it was. Young people, in particular, have such a strong curiosity – they see themselves as global citizens. They're connected to peers all over the world, so the appetite to explore and engage with different cultures is still very much alive.

That said, Covid has introduced a new layer of complexity. Affordability is always a factor, but what I've noticed more is a heightened risk aversion. There's a fear factor at play now that maybe wasn't as pronounced before the pandemic. Some people cite affordability as a reason not to go, but I think it's more about hesitancy and a lack of confidence to take the plunge. Most of our programmes involve earning while you travel, so they're not financially out of reach for many young people – but the fear of the unknown can hold them back.

It's a contradiction, really. On the one hand, you have this generation's curiosity and their desire to embrace the world. On the other hand, there's this lingering anxiety that makes them second-guess whether it's the right time or whether they're ready. As travel providers, I think we play a big role in bridging that gap. It's about offering reassurance and showcasing relatable, inspiring stories from people who've taken the leap before them. That kind of storytelling can be powerful – it helps young travellers see that the challenges are surmountable and that the rewards far outweigh the risks.

Our job is to be relentlessly encouraging, to remind them of the incredible opportunities waiting out there and to provide a safety net of support. When they hear those stories, see the possibilities and feel reassured, they can overcome that fear and rediscover their sense of adventure.

Stephen: I had a conversation with Nick Pound from World Nomads in Chapter 6 and he mentioned that during the initial lockdown period of Covid they had to assist a lot of stranded travellers. Since then, safety and security – both for the travellers themselves and increasingly for their parents – have become significant drivers for young people. Are you finding that participants are asking more questions about these issues now compared to pre-Covid?

Vicki: Absolutely, 100 per cent. We've noticed a clear shift in the types of questions young people are asking. For example, we now have 19- and 20-year-olds inquiring about the medical coverage on their insurance policies. That's a completely fair and reasonable question, of course, but it's not something we would have heard from this age group in previous generations. There's a newfound seriousness about personal safety and financial protection. They're asking questions like, 'How much of this is refundable?' or 'What happens if I have to cancel?' They're taking these matters into account in ways that Millennials or Gen Xers didn't seem to. I think it's a reflection of how this generation is more savvy, more aware and perhaps more cautious than those who came before them.

At the same time – and here's where I start to sound like an old-timer – I think some of the carefree, happy-go-lucky spirit of youth has been lost. There used to be more of a sense of adventure, a willingness to throw caution to the wind and just see how things go. In your twenties, you have so little to lose: no dependents, no major responsibilities and the ability to live on very little money if you need to. You can either take risks, or hurl yourself into a career and work like mad to save for something big. You have options.

I feel like some of that sense of freedom and 'let's just go for it' has disappeared in this generation, and I think that's a shame. Your twenties are the time to be selfish, to do things just for yourself without worrying about long-term consequences. Once you get into later phases of life, like your forties or fifties, you're juggling more responsibilities and have less time and energy to explore in the same way. Youth is the time to embrace those experiences, and I hope we, as an industry, can help remind young people of that.

Stephen: What's interesting – and this isn't something explicitly mentioned in the book but, as you've been talking, it's made me wonder – is whether the accessibility of information now, combined with changes in the cost of living, has shifted the purpose of programmes like Working Holiday visas. Do you think that, compared to 20 years ago, when people might have approached these experiences purely for the adventure, there's now more pressure for them to align with career goals? For example, when I worked in Australia, it happened to tie into my career in hospitality, but a lot of my friends took whatever jobs they could find – gardening, football coaching – and they're now accountants, doctors and so on. Do you think this shift has been influenced by factors like the job market or even post-Covid realities?

Vicki: I think you're spot on. It's so easy for older generations, like ours, to look back and forget how fundamentally different the economic and social environment is for today's young people. When I was entering the job market, there was very little unemployment and the economy felt stable. But if you're part of a generation that grew up seeing your parents struggle through the economic downturn of 2008 and then experienced the uncertainty of Covid first-hand, you're likely to hyper-value stability and security.

That creates a tension when it comes to programmes like Working Holiday visas. On one hand, these opportunities are all about embracing the unknown, leaving your comfort zone, jacking in a job and diving into something completely different. But, on the other hand, this generation has been conditioned by their circumstances to plan every last detail and avoid unnecessary risks. That desire for certainty can sometimes stand in the way of the spontaneity that makes these experiences so transformative.

And yes, the job market has definitely influenced this shift. Affordability is part of it, but I think the need for certainty is even bigger. In today's economic climate, young people feel pressured to justify these experiences as more than just a fun adventure. They need them to contribute to their long-term career goals or feel they're making an 'investment' in themselves. That mindset is practical, but it can also limit the joy and the unexpected discoveries that come from just throwing yourself into an experience for its own sake.

There's never a better time than your twenties to dive headfirst into uncertainty. It's the one phase of life where you have the fewest responsibilities and the most freedom to take risks. The irony is that, by trying to plan every detail, you can sometimes miss out on the most rewarding aspects of travel – the chance encounters, the unplanned adventures and the personal growth that comes from navigating challenges as they arise. Travel, at its core, is about embracing the unknown, and I think part of our role as industry leaders is to encourage young people to reclaim that spirit of adventure.

Stephen: I interviewed Simon Lucey in Chapter 14 and we got onto the topic of how much joy I find in researching a holiday. For me, it's part of the experience. I'll explore Google Maps to find restaurants near the hotel, watch an Anthony Bourdain episode about the destination, or dive into reviews to uncover hidden gems. That dreaming and discovery stage is as thrilling as the trip itself – it builds the anticipation and connection.

Simon agreed, even though he's a bit younger than us. He recently used AI to create a travel itinerary based on his interests and it was surprisingly similar to the one he came up with through his own research.

But here's the thing – what if the younger generation doesn't even know what they're missing in that dreaming stage? Destinations and brands working in youth travel need to think about how they inspire this generation. Whether through TikTok campaigns, striking advertisements or creative guerrilla marketing, the goal has to be rekindling that sense of discovery. The dreaming stage isn't just about planning; it's about igniting curiosity, removing hesitation and encouraging young people to immerse themselves fully in experiences that could reshape their lives.

Vicki: You're right. And I think the key is remembering that AI is only as good as what it knows about you – it starts from your preferences, your likes, your comfort zones. But the best travel experiences, especially in your twenties, often come from stepping out of those boundaries. When I was younger, my plans were often shaped by who I sat next to on a bus. Someone might say, 'You've got to check out Northern Thailand,' and that random suggestion would lead me on an entirely unexpected adventure. AI can't replicate that serendipity or those chance encounters that push you in directions you never anticipated.

The challenge for destinations and travel providers is to showcase this sense of possibility. We need to highlight other people's stories, the unplanned moments and the richness that comes from leaving some gaps in the itinerary. Young people today don't want cookie-cutter experiences, but they do want inspiration and reassurance. That's where peer storytelling becomes so powerful – young travellers don't necessarily want to hear my stories from the Noughties, but they'll resonate with someone their own age who's just had an incredible, unplanned adventure.

It's also a tough sell, because we live in a world that prizes convenience and comfort. But, as we've said, so much of the joy of travel comes from those imperfect moments – getting lost, overcoming challenges and figuring things out. Those are the stories people tell for years. The difficulty is convincing young people that the unpredictable and imperfect parts of travel are often the most rewarding. And the best way to do that is through the authentic voices of those who've been there before.

Stephen: I guess it's about selling the idea that the perfect experience is imperfect. As you said, it's hard to do, but one brand that managed it brilliantly was Hans Brinker in Amsterdam. When I interrailed, I stayed there because of their marketing. They leaned so far into their imperfections – they even framed complaint letters in the reception area. It was as bad as they said it would be, but that was the charm. It was a genius way of marketing, so much so that they've won countless awards. They had a brilliant ad agency that's since helped rebrand many travel and hospitality businesses.

Vicki: Exactly. Leaning into imperfections is a bold move, but when it's done well, it works. People want authenticity – they want to know what they're signing up for is real.

We've used a tagline for summer camps in the past: 'The hardest job you'll ever love.' I don't think we're using it right now, but it perfectly captures the essence of what working at a camp is like. You're working long hours with kids every day. It's tough – some people leave in the first week because they realize it's harder than they imagined. But the ones who stay never regret it. In fact, camp tends to create lifelong bonds. We hear stories all the time of people meeting their best friends or even their future spouses at camp. Campers and counsellors often stay in touch for years and the experience is universally life-changing.

It's actually our highest satisfaction programme and yet, it's also one of the hardest. So yes, it's a tough sell. But there's still such a strong demand. Young people are still eager to step outside their comfort zones. Sometimes, there's a perception that this generation is entitled or too comfortable, but they know better. They understand the value of pushing themselves, of experiencing something transformative. What we have to do is address the fears – not just of the young people but of their parents too.

Stephen: We've touched on some of the barriers for young people, but as a business operating in youth mobility – not just since Covid but throughout your entire existence – what are the biggest challenges you've faced in getting young people moving around the world?

Vicki: There are two major barriers we consistently encounter. The first is one we've already touched on: young people are constantly evolving, and so are their wants and needs. Their preferences shift rapidly, and as a business we have to decide whether to push back against those changes or meet them where they are. It's not an easy decision, but I'd say we focus on evolving with them.

The pace of change is incredibly fast, and staying relevant requires us to hire a diverse team of talented individuals across different age demographics. It's about understanding what young people want today while also anticipating their future needs – even the ones they might not realize they have yet. It's a delicate balance, but one that's critical to our success.

The second and by far the bigger challenge is preserving these programmes from a visa and youth mobility standpoint. Globally, there's an increasing trend towards tightening borders and limiting the free movement of people. This has created immense pressure on programmes like ours. Fortunately, there's also a strong opposing faction that understands the immense value of youth mobility, but it's a constant battle.

We can never stop promoting and advocating for these programmes. The reality is that government policies and personnel change frequently, and with every new administration or shift in priorities we often find ourselves starting from scratch. It's not uncommon to have the same conversation with different government departments every six months because the people in charge don't always understand the nuances or long-term value of these programmes. That's why continuity and persistence are so important.

Brexit is a prime example. I genuinely believe the negative impact on youth mobility wasn't an intentional decision but rather an unintended consequence. Regardless, it's been a real consequence. UK young people have lost the ability to work and travel in Europe in the same way they once could, and that's a significant loss for them – and for Europe. It underscores how vital it is to preserve these programmes and their frameworks.

So, to sum up, I'd say our two biggest ongoing challenges are keeping up with the ever-changing needs of young people and safeguarding the long-term future of these mobility programmes. Both are crucial and require constant attention and advocacy.

QUESTION

If you want to attract a younger audience to your brand or destination have you looked at hiring people that are the age of the target market to make sure your product and marketing are relevant?

Stephen: And I guess your success helps sustain the entire ecosystem of youth travel, which you're obviously very engaged with. From hostels and adventure sports to insurance providers and even the local economies where these young people work, it's all interconnected. So, I imagine it's vital for you to keep people moving, not just for your organization but for the health of this broader system.

Vicki: What we do has a ripple effect across the entire youth travel sector. Many of the people we help send overseas end up engaging with other industries or roles that keep this ecosystem thriving. Whether it's working in a hostel, participating in local adventure activities or contributing to the local economy by simply living and working there, it all adds up to a significant impact.

And, honestly, I don't want to sound too idealistic but I do believe in the greater value of these programmes. There's something profoundly important about the cultural exchange and understanding that comes from youth travel. It's not just about work or play – it's about fostering dialogue, breaking down barriers and building bridges between different cultures. In today's political climate, that's more crucial than ever. It's about more than economics; it's about fostering empathy and global citizenship. That might sound like a beauty pageant answer, but it's true. These programmes really do have the power to promote mutual understanding, which is a foundation for peace and progress.

One more thing I've reflected on, particularly as I attend conferences and talk to colleagues in the sector, is the perception of young people. Some of us in the industry have been doing this for a long time and it's easy to fall into the trap of framing young people's evolving preferences and behaviours in a negative light – things like calling them entitled or saying they have unrealistic expectations. But, honestly, I think every generation gets labelled like that by the one before it. I'm sure our parents thought the same about us.

Instead of criticizing, we should see these changes as opportunities. This generation is shaping the future, driving progress in ways we might not fully understand yet. As professionals in youth travel, we can't afford to resist or lament the ways they're different. We need to adapt, embrace the unknown and work with their evolving values and expectations. After all, they're not just our customers – they're our industry's next generation of leaders and innovators.

Stephen: In reality, you're moving thousands of young people around the world every year, but you're not as large as some of the biggest travel brands or destinations that deal with millions. That nimbleness, that ability to adapt to shifting trends, becomes so essential for destinations and brands to effectively engage with youth. If they rest on their laurels or try to rely on the same strategies or products from 10, 15 or even 20 years ago, they simply won't connect. This generation's expectations and needs have evolved so much. Take hostels, for example. The hostels we stayed in 20 years ago wouldn't meet the demands of today's travellers at all.

Vicki: That's such an important point. If I can use Canada as an example, I think it serves as an interesting case study of how a government can proactively engage with the youth travel market. Canada has developed a government department specifically focused on this, but they recognize their limitations – they don't have direct, day-to-day interaction with the young people coming in and out of the country. So, instead, they've partnered with eight or nine key organizations in the youth travel sector.

What's brilliant about this collaboration is that these organizations provide the government with critical data, trends and insights about youth travellers. They work together on research and use that information to refine and adapt their programmes. It's a neat example of a government recognizing its gaps and addressing them in a way that creates a robust, forward-thinking programme.

Canada's approach shows how valuable it is when governments actively interact with the youth travel sector. It leads to a more dynamic and resilient programme that's in tune with the needs of today's young travellers. It's a model that I think other destinations could learn a lot from. When governments recognize that collaboration with private sector experts can fill in the gaps in their knowledge or reach, the results can be transformative. It's a real testament to the power of working together.

Conclusion

As this chapter demonstrates, youth mobility is far more than an opportunity for adventure. It is a foundational experience that cultivates resilience, adaptability and a global mindset in young people. These qualities not only shape the travellers themselves but also have a lasting impact on the destinations they visit and the industries they engage with.

Vicki's insights illustrate the delicate balance of challenges and opportunities in the youth mobility space. The evolving preferences of young travellers require constant adaptation and innovation, while the preservation of mobility programmes in the face of political and regulatory challenges demands relentless advocacy. Yet, the rewards of these efforts are undeniable – destinations and brands that embrace youth travel are investing in future leaders, loyal visitors and economic contributors.

The power of youth travel lies not just in the short-term economic benefits but in the lifelong connections and cultural exchange it fosters. Whether through Working Holiday visas, internships or cultural exchange programmes, these experiences are life-changing for individuals and transformative for societies.

KEY TAKEAWAYS

1   **The transformative power of youth travel:** Youth travel fosters resilience, open-mindedness and problem-solving skills in young people. These experiences empower individuals to navigate challenges, embrace uncertainty and develop a global perspective.

2   **Economic and cultural impact:** Youth travellers contribute significantly to local economies by staying longer, exploring diverse regions and participating in work-and-travel programmes. Their influence extends beyond their immediate presence, as they often inspire friends, family and peers to visit the destinations they've explored.

3   **Youth mobility as a pathway to leadership:** Programmes like BUNAC's are not just opportunities for exploration – they are incubators for leadership. The skills and perspectives gained through youth travel often translate into entrepreneurial ventures, managerial roles and even global leadership positions.

4   **The role of collaboration and policy:** Effective youth mobility programmes rely on collaboration between governments, organizations and private-sector players. Destinations like Canada exemplify how partnerships and data-driven insights can create robust and adaptable programmes.

5   **The importance of authenticity and imperfection:** Authentic experiences, including the challenges and imperfections of travel, are what make youth mobility transformative. Destinations and brands that lean into these realities and communicate them effectively can create deeper connections with young travellers.

6 **Overcoming barriers:** The biggest challenges in youth mobility include adapting to the evolving preferences of young people and preserving visa and mobility frameworks in an era of increasing restrictions. Advocacy and innovation are essential to overcoming these obstacles.

7 **The long-term value of youth travellers:** Youth travel is not just about the immediate economic impact; it creates lifelong connections and loyalty. Investing in young travellers today leads to long-term benefits as they return as business travellers, family vacationers and even luxury tourists.

12

Reinventing hostels for a new generation

Challenges and opportunities

In this chapter we delve into the perspectives of Paloma Meng, General Manager of Shanghai and Head of Supply for Asia-Pacific (APAC) at Hostelworld Group, a leading distributor of hostel accommodation favoured by young travellers. Paloma provides valuable insights into current travel trends in the APAC region, highlighting how young people from Asia are exploring the world and how international youth are engaging with Asia as a travel destination.

Given the vast population and diverse cultures across the region, Paloma's expertise sheds light on the evolving dynamics of youth travel, revealing shifts in behaviour, preferences and opportunities for growth. Her insights offer a comprehensive look at where the travel industry should focus its efforts to cater to this vibrant and influential market segment.

Stephen: Paloma, thank you for joining us today. Can you start by giving us a bit of background on your role at Hostelworld and what your team focuses on?

Paloma: Of course. I'm based in Shanghai and have been working with Hostelworld for over a decade. Hostelworld has had a presence in Shanghai for more than 18 years and our office primarily focuses on the supply side – sourcing hostels and budget accommodation throughout the APAC region.

Hostelworld's role in the evolving youth travel market

Stephen: That's impressive. Hostelworld has become synonymous with youth travel, especially among budget-conscious travellers. Can you give us a brief overview of what Hostelworld offers?

Paloma: Hostelworld is one of the leading online platforms for booking hostels and budget accommodation globally. We have over 17,000 properties in around 180 countries and territories. Our platform caters to a wide range of travellers, including young backpackers, solo adventurers and budget-conscious groups who are looking for affordable yet authentic travel experiences.

Stephen: It sounds like a vibrant market. How has Hostelworld adapted to the evolving travel trends, especially in such a diverse region as APAC?

Paloma: The APAC market is incredibly dynamic, with young travellers constantly seeking new experiences and destinations. Our focus has been on understanding these evolving preferences and ensuring that our supply of hostels and budget accommodation meets the needs of this diverse audience. It's about connecting young travellers with unique, affordable options that enhance their travel experience.

Stephen: As someone who stayed in hostels globally – though many years ago – I've noticed how much they've evolved, especially in the last decade. Traditionally, hostels were geared towards younger travellers. Have you observed a shift in demographics, or are they still predominantly chosen by younger people?

Paloma: Hostels have always been synonymous with youth travel, but we've seen a shift in demographics in recent years. Previously, the typical age range for hostel guests was between 18 and 27, but now it's expanding into the mid-30s. This change reflects broader trends among Millennials and Gen Z, who increasingly prioritize experiences over material possessions. As a result, they seek out affordable accommodation options, like hostels, which allow them to allocate more of their budget to activities and adventures.

In response to these changing preferences, we introduced a feature called 'Linkups' on our app. This feature promotes events and activities within hostels before guests even arrive, enhancing the social and experiential aspect of their stay.

The changing face of hostel guests

Stephen: With the changing demographics of hostel guests, how have hostels adapted to meet these new demands? In one part of this book, we attempted to define youth travel as ranging from 11 to 35 years old, with 11- to 18-year-olds travelling independently of their parents but within organized groups and 18- to 35-year-olds often travelling independently or in smaller groups. When I was backpacking between the ages of 18 and 23, hostels were geared primarily towards that age range. Since then, the facilities and types of rooms on offer have changed significantly. Can you elaborate on how the hostel product has evolved over the years?

Paloma: Hostels have adapted tremendously to meet the needs of a broader range of travellers. Many hostels now offer private rooms and ensuite bathrooms, which cater not only to younger travellers but also to older guests seeking more privacy and comfort. There's been a diversification within the hostel category itself; some properties market themselves as 'backpacker hostels', while others have rebranded as 'poshtels' offering a more upscale experience.

We're also seeing a rise in co-living spaces within hostels, which cater to digital nomads who want a blend of travel and work. These properties often feature coworking spaces and even private offices, allowing travellers to use the hostel as a base for both leisure and professional activities. This evolution reflects hostels' flexibility – they can quickly adapt to changing traveller preferences because they are often independently owned.

Additionally, hostels have significantly upgraded their facilities, with stylish common areas, comfortable beds and high-speed Wi-Fi now standard. The emphasis on enhancing the overall experience has changed the quality of hostels dramatically compared to 20 years ago, making them a far more appealing choice for a wider demographic.

QUESTION

The quality of hostels has changed but sometimes senior decision-makers do not realize that. What hostels are available in your region, and have you taken a look at them recently?

APAC's growing hostel landscape

Stephen: Earlier in the book, I shared some of my own travel experiences, including a memorable trip through Vietnam in 2003. I booked my first night in a guest house for just $2 a night using Hostelworld. You're based in the APAC region, where the youth travel ecosystem is still thriving, but I imagine the guest accommodation landscape has changed significantly over the past 20 years. While the essence of travel remains the same, the overnight experience seems vastly different now. Could you outline how accommodation has evolved across the region?

Paloma: Absolutely, the accommodation landscape in popular backpacker destinations like Thailand, Vietnam, Indonesia and Australia has evolved significantly. We're also seeing growing interest in places like Japan and the Philippines. The supply of high-quality hostels in Asia has increased dramatically, but there are notable differences compared to other regions like Australasia.

One key difference is the scale. Hostels in Asia tend to be much smaller, often about half the size of those in Australia, primarily because average bed prices in Asia are significantly lower than in Europe or Australasia. Many hostels are privately owned and while they may be smaller, they're often tailored to meet the needs of modern travellers.

We've also seen the rise of high-quality hostel brands emerging from the region, such as Lokti, which markets itself as a 'poshtel'. These properties blend hotel-standard accommodation with the social, budget-friendly vibe that appeals to young travellers. This shift reflects the broader trend of elevating the hostel experience to meet changing expectations without losing the essence of affordable and social travel.

The rise of digital-first travel

Stephen: Back when I travelled, I relied on a guidebook and smartphones didn't exist. We used internet cafes to send emails home or research train and plane schedules, but mostly it was about showing up and finding places to stay on the spot. I imagine booking patterns have changed dramatically and today, attracting young travellers must involve being mobile-first, right?

Paloma: Definitely. In 2003, there were far fewer hostels featured on our platform in Asia, and even guest houses were challenging to vet for safety and cleanliness. Hostels as an accommodation type were not well established in Asia back in the early 2000s. This began to change significantly around 2010 when the market recognized the value of investing in hostels and catering to young travellers. We started seeing more investment in larger, European- or Australian-style hostels, especially in places like Thailand, which then spread across the region.

Asia is a top destination for young travellers, so creating the right type of accommodation is crucial for tourism growth in these areas. Japan, for instance, began investing heavily in hostels around 2015 and, despite the challenges posed by Covid, there's still a strong commitment to fostering inbound youth travel, especially in Japan.

Although Asia's hostel scene developed later than Australia and Europe, many of the hostels in the region now rival the quality of small independent budget hotels in Europe. In warmer destinations we've also seen the emergence of resort-style hostels that offer a unique mix of social and high-quality amenities. This variety in hostel offerings caters to the evolving needs of young travellers who, in turn, share their experiences widely on social media, further promoting these destinations among their peers.

Understanding Asian traveller preferences

Stephen: When I travelled, experiencing authentic local food and culture was incredibly important to me. During my three months in Southeast Asia, I felt immersed in the local vibe, even in the guesthouses where we stayed. However, in Australia, while the hostels were cleaner and more modern, they lacked that same local feel. The experience was more people-focused than place-focused, which was still fantastic but different. I suppose that the desires of travellers vary depending on their age, not just in terms of accommodation but also in the experiences they seek. Have you noticed any shifts in travel patterns within Asia, especially post-Covid, or over the last decade? Additionally, have there been changes in how young Asian travellers approach their travels and where are the current hotspots in Asia?

Paloma: The Asia-Pacific region continues to see growing demand, especially when compared to Europe. This trend applies to both hotels and hostels. Southeast Asia, in particular, has experienced significant growth, largely fuelled by the rise of low-cost airlines that connect young travellers across the region. Combined with relatively low accommodation prices compared to Europe, Australia and the US, Southeast Asia has become a hub for youth travel.

Young travellers often stay longer in Southeast Asia than luxury travellers and explore more remote areas, which positively impacts local economies beyond the usual tourist spots. Recently, there has been increased interest in Northern Asia, with Japan becoming particularly popular. However, traditional destinations like Thailand, Vietnam, Indonesia and the Philippines remain at the top of the list. In the broader APAC region, Australia remains a staple destination for backpackers, with a well-established youth travel industry that predates much of Asia.

For European and English-speaking backpackers, there's a noticeable shift towards spending more time in Asia. This shift is largely due to the lower costs compared to traditional Western destinations and the improved infrastructure brought about by low-cost carriers, which make travelling across the region easier and more affordable. Social media has also played a huge role in promoting destinations like Bali, Japan and Thailand, driving interest among younger travellers.

For young Asian travellers, their preferences often differ significantly from their Western counterparts. Asian travellers typically prefer private rooms over dormitories and are willing to pay more for privacy. They prioritize cleanliness, safety and quiet accommodation, avoiding places with a big party scene. Security is also a major concern, not just in terms of location but also within the accommodation itself.

Cultural preferences also vary greatly across Asia due to diverse languages, social dynamics and payment habits. For instance, young travellers from Malaysia and Indonesia often prefer to prepay for everything before arriving at their destination. In contrast, Chinese travellers primarily use payment methods like Alipay and WeChat, rather than Western credit cards. Therefore, it's crucial for accommodation providers to offer these payment options to attract young Chinese travellers, making the booking process more convenient and aligned with their preferences.

Korean travellers have distinct preferences as well. Many young Koreans are drawn to Australia, particularly for backpacking or working holiday experiences. This influx of Korean travellers is welcomed by Australian hostels, which are benefitting from the growing number of guests from Korea. The cultural exchange and the long stays often associated with working holidays make Korean travellers a valuable demographic for the Australian hostel industry.

QUESTION

Is your brand or region equipped to capture the Asian market? Do you have the right payment methods and other facilities to make the traveller feel at home?

Stephen: That's quite different from when I travelled. Back then, we mostly dealt in cash, transferring money at borders and relying on travellers' cheques. I remember feeling like a millionaire in Vietnam when I exchanged money and suddenly had a million Vietnamese dongs! It's clear things have evolved significantly since then.

Today, European travellers have digital banking options like Revolut, which offer free foreign currency usage and make managing money abroad much easier. Even the smallest guesthouses need to keep up with these digital payment trends, not just in Europe but globally. For Asian travellers visiting Europe and vice versa, it's crucial that accommodation providers are equipped to handle modern payment methods, so travellers don't have to worry about currency exchange or cash management. It's all about seamless travel experiences now.

Leveraging marketing to target Asian travellers

Stephen: Today, the phone is undoubtedly the most important device when travelling, especially when you're remote. This reliance has only intensified since Covid, where contactless payments became the norm in many countries. Before the pandemic, phones were primarily used for boarding passes and banking apps, while most people still carried physical cards. Now, it's incredible to see how even traditionally cash-only places in Europe have adapted to card payments. Ensuring this payment flexibility is crucial, particularly in the youth market, where phones – and increasingly smartwatches – are permanently connected to young travellers.

I remember running a hotel in the late 2000s when there was a big push in the UK to attract Chinese tourists. We were told we needed to adapt our offerings, from adding Chinese breakfasts to hiring Mandarin speakers at the reception. However, we soon realized that those travellers were mostly large groups or luxury travellers, not the individual young traveller we had initially hoped to attract.

So, if a brand or destination is now trying to appeal to young travellers from Asia, whether from China or Southeast Asia, what would you advise from a marketing and product perspective?

Paloma: That's a great question. Let's break it down by different groups: Japanese, Korean, Chinese and Thai travellers.

Starting with Japan, the current weak yen has significantly reduced the number of Japanese travellers, especially among young people. I recently spoke with a hostel owner in Japan who expressed concerns about youth travel. Less than 25 per cent of young Japanese even have a passport, making staycations the norm rather than travelling abroad. There's a need for something to inspire young Japanese travellers to step out of their comfort zones and explore internationally.

In contrast, Korean travellers are more adventurous and are heavily influenced by social media, celebrities and TV shows. If a reality show features a particular destination, you'll see an immediate spike in interest among young Koreans. For the hostel industry, Koreans are still one of the more promising Asian markets as they are more open to using hostel facilities and adapting to the culture.

Chinese travellers are quite diverse and we need to distinguish between different segments. There are domestic travellers within China, first-time or second-time outbound travellers and Chinese youth who study or live abroad. For Chinese students overseas, hostels are often a preferred accommodation option. However, domestic young travellers in China have a different understanding of hostels, often using them as temporary, low-cost accommodation while job hunting in big cities. The concept of hostels as social, travel-oriented accommodation is less familiar to them.

Lastly, Thai travellers typically prefer to travel in small groups, whether with family or friends and they often choose hostels for their cost-effectiveness compared to hotels. They usually book private or family rooms, making hostels an attractive option due to the lower costs while still offering a social experience.

Stephen: Have you noticed any shifts in European and Australian hostels to better cater to the Asian market? Or are these hostels still primarily designed with European and US travellers in mind, with Asian guests fitting in where they can? If they don't fit, do they simply opt for hotels?

Paloma: I don't see a major shift yet. In key European markets, hostels generally have high occupancy rates and cater to a mix of nationalities simply because demand is so strong. Since everyone books through platforms in advance, the average daily rate is quite high and hostels don't necessarily prioritize tailoring their offerings to specific nationalities. It's more about maintaining a universal appeal.

In Australia, I did see some of the bigger hostels actively looking to attract more travellers from Asia, particularly from Korea and China. However, I wouldn't say there's a clear, dedicated strategy across the board to cater specifically to the Asian market.

Stephen: If I were setting up a new hostel, perhaps in a secondary city in Europe, and wanted to target the youth market from Asia – let's say Southeast Asia, Korea and China – what should I consider? Should I focus on communal spaces with amenities like wok burners to make Asian travellers feel more at home, or should I emphasize local cuisine to give them a taste of the local culture?

Paloma: First, it's important to understand that demand is not something you can fully control. If you're setting up a hostel in a secondary European city, it's crucial to identify the demographic trends of visitors to that location. If there's significant growth in Asian travellers, then catering to that market makes sense, regardless of whether you're running a hotel or a hostel.

You'll need a clear rate and distribution strategy tailored to where your demand is coming from, whether that's Hostelworld or other online travel agencies. It's also essential to have a strong presence on these platforms, with proper property profiles. This means having your content translated into relevant languages, optimized with the right search terms and set up with effective pay-per-click and search engine optimization strategies.

While having your own app might be challenging for independent hostels, a mobile-friendly website in multiple languages can make a big difference in your search visibility. Translated content on online travel agencies is also vital for reaching your target markets.

When it comes to catering specifically to Asian travellers, consider making simple, culturally familiar enhancements. Offering an Asian-friendly breakfast or providing a wok in the shared kitchen, along with essential ingredients like soy sauce, can make your hostel feel more welcoming. It's about small touches that show you're accommodating different cultures.

Since many Asian cultures are more reserved compared to European ones, hosting simple networking or food-tasting events can create a warm, inclusive atmosphere. It's not that these travellers don't want to meet others; it's just that the social dynamics are different. Treating them like friends or family can encourage positive word-of-mouth, especially among Korean travellers, who love to share their experiences online. Once something goes viral on their social media, you can expect a significant influx of young travellers from that market.

For Chinese travellers, the key is to understand their unique digital landscape. One particularly popular platform among young Chinese female travellers is Xiaohongshu, also known as Little Red Book. This app is often the go-to place for travel recommendations and reviews. If your hostel or brand can gain popularity on Little Red Book, it's akin to going viral on Instagram outside of China. TikTok also works well, though WeChat's reach is more limited due to its closed nature.

For brands looking to attract Chinese female travellers, it's highly recommended to invest time in understanding platforms like Little Red Book and getting your properties featured there. This targeted approach can significantly boost your visibility among that demographic.

TIP

When looking at targeting the Asian traveller with marketing campaigns, it is unlikely a blanket approach will work. Work out which country you are focusing on and research that country's nuances – whether it be where to market or what the expectations are from that traveller.

Stephen: This really highlights the challenges and opportunities that have emerged compared to when I travelled 20 years ago. Back then, without social media, word-of-mouth referrals only happened after I returned home from my trips. Now, brands, destinations and hostels can become part of the travel story in real-time, much earlier in the traveller's journey.

If the product has the right look and feel, along with targeted services, translations and cultural touches, it can quickly gain traction. For secondary or even tertiary cities, especially those that have featured in popular Korean shows, there's a unique opportunity. By combining targeted marketing with the right product offerings, flexible payment terms and easy access, these locations can attract a surge of young travellers.

However, it's not just about spending on marketing – you need to ensure the investment aligns with these key factors to truly see a return.

Looking at the global landscape, are there any specific trends you've noticed in travel and accommodation? Are there any hot destinations that are emerging globally, or notable shifts within the accommodation space? For example, in the airline industry, we've seen business class offerings evolve dramatically over the past decade – from open-plan seating to private, cocoon-like pods. Have you observed similar changes within the hostel or broader accommodation sectors?

Paloma: Globally, we've seen a significant rise in travel to Asia post-Covid, driven by factors like affordability, ease of access and the ability to visit multiple countries on one trip at a lower cost compared to Europe or the US.

Europe remains a traditional favourite for backpackers, with Spain, Italy and France consistently popular among young travellers. Latin America also saw a surge in popularity during and after Covid, while Australia continues to be a major backpacking destination, especially for those on Working Holiday visas who often stay for several weeks or months.

These destinations remain key for our market. But beyond the popular spots, there's been a shift in how technology is enhancing the travel experience, particularly for youth travellers. At Hostelworld, our mission is to connect people and we've developed features that foster community among travellers even before they arrive at their destination.

One of our recent innovations is the Hostelworld Chat feature. Fourteen days before check-in, customers are automatically added to a city-level chat room, allowing them to connect with others heading to the same destination. For example, if you book a stay in Barcelona, you'll be added to the Barcelona chat room, where you can chat with other travellers. This chat remains open until three days after check-out, giving travellers the chance to decide if they want to meet others, attend events, or explore activities together. If a traveller wishes to leave the chat, they can do so anytime.

Another feature we've introduced is called Linkups, which highlights hostel events and activities within a city. For instance, if you're staying at Wake Up! Hostel in Sydney, you can see what events are happening there, or even explore activities at other hostels in the area. You can book directly through the app, whether it's a party at another hostel or an adventure activity, allowing you to plan your journey and make connections along the way.

These technological innovations are about creating a social, community-driven experience for travellers, making their journey more engaging and interactive. It's about connecting people, fostering new friendships and enhancing the overall travel experience.

Balancing technology and authentic experiences

Stephen: This is a significant shift in travel, and it's clearly been challenging for brands and destinations to keep pace with rapidly changing trends and technological advancements. As you develop these digital products, I imagine you'll see varying levels of engagement among different nationalities and in different cities. Larger cities might draw more interaction simply because they can feel overwhelming, and for young travellers from big cities visiting another urban hub can be the easiest way to connect with others.

Traditionally, these connections happened at hostel bars, in shared kitchens, or over breakfast. Now, with technology facilitating these interactions, there's a new, more connected way of meeting people, though it does present a challenge. With so much information available, creating those 'wow' moments for travellers has become more difficult – surprises are fewer and experiences are often anticipated rather than discovered.

When I travelled to Cambodia in 2003, not much was known or written about the country due to its recent history with the Khmer Rouge. Seeing Angkor Wat at sunrise was a powerful, unexpected moment that still gives me goosebumps. Part of the magic was the sheer surprise – I couldn't believe I knew so little about Cambodia, and that sense of discovery was a big reason why I travelled in the first place.

Today, young travellers are hyper-connected and often know almost everything about a destination before they even arrive. This poses a significant challenge for brands, hostels and destinations: how can they still deliver those unforgettable 'wow' moments when so much is already known?

Despite our tendency to say we don't like surprises, I believe we actually crave them, especially those moments that touch our emotions and senses in unexpected ways.

Paloma: You raise an excellent point. When we think back to the days when we printed maps and used guidebooks to find hostels, it was a very different travel experience. Now, everything is tech-driven, with information available at our fingertips. This explosion of accessible information isn't just in travel – it's in every industry, and it often makes everyone feel like an expert even before they leave home.

Authenticity is something travellers crave, but it's increasingly missing from many experiences. At Hostelworld, our initial goal in launching our social platform was to help solo travellers find a sense of community and avoid feeling alone in unfamiliar destinations. That was the core strategy behind creating these chat functions and pre-arrival connections.

We do recognize that this approach can alter the traditional charm of discovering a destination on your own. It depends on how much information is shared and the nature of those pre-arrival chats and events. However, so far, the feedback has been overwhelmingly positive. Young and solo travellers often tell us that these features add an element of fun and anticipation rather than making them feel as though they know everything and have lost interest.

The broader travel industry is grappling with this challenge – maintaining a sense of authenticity and delivering those special 'wow' moments. It's a balancing act, and we need to continue finding ways to keep travel exciting and emotionally engaging, despite the abundance of information available.

Mental health and travel: Disconnecting to reconnect

Stephen: It's definitely a challenge, whether in education or travel, to keep young people engaged, because their attention spans are much shorter these days. There used to be many moments in a day when we were naturally disconnected from technology, simply because it wasn't as readily available. Now, with phones and smart devices, we are constantly connected.

Personally, I think there will come a time when people will want to disconnect more. While having everything at your fingertips is convenient, it also creates problems. There's been a lot of research and discussion on the impact of smart technology on mental health, particularly among younger people. I wouldn't be surprised if, one day, people react to tech the same way they did when they realized cigarettes were harmful.

Travel offers a great way to counterbalance this, improving mental health by encouraging disconnection. I hope young people will find that, by unplugging while travelling, their mental wellbeing improves, allowing them to use technology to enhance their experiences rather than letting it control them.

I know some people who have turned off many notifications on their phones because they felt overwhelmed by the constant alerts and bad news. They've found that, by changing how they interact with their devices, they've become much less anxious.

What we often see now, especially among younger age groups, is that they engage with platforms like X (formerly Twitter) for quick snippets of information rather than reading full articles. While this trend can lead to incomplete understanding, it reflects the reality of media consumption today. It's an interesting trend, and it's definitely something that everyone in the travel industry needs to keep a close eye on.

Paloma: You're right, and it reflects a broader societal shift. I recently listened to an interview with Joan Chen, a well-known actress who starred in *The Last Emperor*. She spoke about her daughters and pointed out that the current generation struggles to focus, largely because of digital devices and constant interruptions.

Young people today can't stand being bored. For example, many would find an eight-hour bus ride unbearable because they can't access their devices. But back when we travelled, those long rides were moments of quiet reflection, a time to meditate, think and simply enjoy the scenery. It's clear that technology has dramatically changed our behaviour compared to the past.

This constant connection to devices and the flood of notifications can be exhausting, which is why so many people feel the need to switch off, step back and even find solace in reading a book. As technology continues to develop, I think it's important for us as a society to be mindful and deliberate about how we manage our time and lives, ensuring we don't lose the ability to disconnect and enjoy simple moments.

TIP

Balancing digital connectivity with real-world immersion is essential for engaging young travellers effectively. The travel sector needs to monitor how young people's interactions evolve, ensuring that while they remain connected, they also experience the destination and brand meaningfully. Additionally, with mental health becoming a prominent concern among young people, the industry must consider this factor in designing travel experiences that support both engagement and wellbeing.

Stephen: I'm quite attached to my phone as well, mostly for work and keeping up with travel and food inspiration. But I'm aware of how it can affect me – I sometimes feel like I can't even go to sleep without checking my phone, even though we all know that reading on a bright screen isn't great, especially because of the blue light.

This has also become an issue in shared spaces like dorms in hostels, where some people are on their phones late into the night while others are trying to sleep. It's another example of how our attachment to technology is starting to impact not just individual experiences but shared ones as well.

Paloma: That's an interesting point. I recently met Howard in Malacca, who runs a well-known social hostel called Ringo's Foyer. During a conversation in Chennai in May, he shared his concerns with me. He was concerned as to what was going on with backpackers these days and felt that young travellers don't seem to want to go out or socialize in the common areas anymore – they just stay on their phones.

Howard's hostel is all about creating social experiences. He takes guests to local markets, introduces them to local food and offers city walks. But he's noticing a shift in behaviour; guests seem less inclined to participate. He's worried about how to motivate them to leave their beds and engage during their stay. It's a real challenge when the essence of a social hostel is being eroded by changing habits.

Stephen: It's been fantastic speaking with you and I've learnt a lot from our conversation. There are clearly significant changes happening within the accommodation side of youth travel, but with these changes also come great opportunities. Hostels that stay attuned to these trends and understand their target market have a huge opportunity to succeed. By delivering authentic experiences that resonate with young travellers, they can truly stand out in a crowded market.

Conclusion

Paloma emphasizes that, while technology has transformed how young people travel, there's a vital need to balance digital convenience with authentic, memorable experiences. Hostelworld is adapting by incorporating social features that enhance connectivity among travellers without overshadowing the essence of exploration and discovery. The travel industry must continue to evolve, focusing on personalized experiences, cultural understanding and opportunities for travellers to engage meaningfully with destinations and each other.

KEY TAKEAWAYS

1 **Expanded demographics in hostels:** Hostels now attract a wider age range, including travellers into their mid-30s, focusing on experience-rich stays. Modern hostels offer amenities like private rooms, ensuites and co-living spaces, blending social atmospheres with upscale features ('poshtels').

2 **Growth in the APAC hostel market:** Significant investments in hostel infrastructure have elevated standards in Southeast Asia, Japan and Australia, with high-quality brands like Lokti emerging.

3 **Technological impact on travel:** Digital platforms and mobile-first strategies have reshaped booking patterns and traveller behaviour. Features like Hostelworld's Linkups enhance social interaction pre-arrival.

4 **Diverse preferences of Asian travellers:**

   o Chinese travellers prefer private accommodation and use digital payment methods. Travel inspiration often comes from Xiaohongshu.

- o Korean travellers are adventurous and social media-influenced, with a high interest in Australian travel.

- o Japanese travellers are often domestic-focused with lower passport ownership, requiring inspiration for international trips.

- o Thai travellers favour small-group travel and affordable private rooms.

5 **Marketing and cultural adaptation:** Successful hostels provide culturally relevant amenities and content, translating property profiles into native languages and utilizing local social media and payment platforms.

6 **Challenges of over-connectivity:** Excessive use of technology may reduce authentic experiences, with hostels facing difficulties fostering social interaction due to guests' device attachment.

7 **Mental health and travel:** The always-connected lifestyle impacts mental wellbeing, but travel offers a chance for young people to disconnect and rejuvenate.

8 **Opportunities for hostels and destinations:** By understanding trends and adapting, hostels can differentiate themselves, offering authentic, culturally sensitive experiences that appeal to young travellers seeking community and genuine interactions.

13

Redefining youth living

From hostels to co-living

The evolution of youth-focused accommodation – spanning hostels, purpose-built student accommodation (PBSA) and co-living – reflects the shifting priorities of a dynamic, global generation. In this interview with Neil Smith of Scape, we delve into how the accommodation landscape for young people has matured over the past two decades. Neil, whose experience spans hostels and large-scale student housing developments, provides insights into how youth travel and housing needs have evolved. From health and wellbeing to flexibility and community, this interview explores the trends shaping PBSA, the impact of Covid and the emerging role of co-living as a bridge between student life and the professional world.

Introduction to youth-focused living

Stephen: Neil, you work for a company called Scape. Could you explain what the organization does, where it's located and the different brands under its umbrella?

Neil: Of course, Stephen. Scape is a student living specialist company. While we initially focused on the student sector, our reach has since extended with the addition of our residential brand called Morro. While Scape caters specifically for students, Morro caters for anyone who is looking for flexible accommodation with high-end amenities and a sense of community. We prefer not to use the term 'co-living' as it doesn't fully capture what Morro is about. However, it does provide a helpful frame of reference for the purpose of this interview.

We currently operate Scape in Australia and the UK and Morro in the UK and the US. Scape started in the UK about 12 years ago with a simple yet ambitious idea: students deserve better. We wanted to create a product that was designed from the inside out, truly meeting the needs of students. It was about creating rooms and spaces that were not just functional but high-quality and intelligently designed, supportive of community and the student experience.

We launched our first Scape property in London with 588 beds. Since then, we've expanded to about 4,500 beds in the UK, more than 20,000 beds in Australia and we're building a growing portfolio in the US. We're also exploring opportunities in Europe.

In the UK, we've been operating for over a decade now. While we're predominantly focused on London, we've also ventured into other locations such as Leeds and Guildford near key Russell Group university cities. It's been a successful journey so far, but we still consider ourselves in the early stages with plenty of room for growth and evolution.

TIP

Do you have a defined target market of young people? Rather than a broad-brush approach why not find a smaller target market that fits best with your product.

Stephen: Cool. It's fascinating to hear about your background, Neil. I think we first met in the youth travel sector, probably 10 or 15 years ago, when you were managing a hostel in Australia. I'd love to hear your perspective on how accommodation for young people has evolved over the years. From your experience with hostels in Australia to PBSA in the UK and across Europe, how have the demands of young people changed?

Neil: That's right – it feels like a long time ago now! I began my career in Australia and later moved to the UK, where I got involved with the rollout of the Generator Hostels brand across Europe in destinations including Sweden, Italy, France and the Netherlands. That took me to a lot of different city centres, which gave me great insight into how accommodation evolves across markets and demographics.

Looking at the progression of that product and then comparing it to PBSA, there are definitely some commonalities, as well as significant differences between the two audiences – primarily due to life stages and what they're seeking from their accommodation.

One thing I've noticed across both sectors is the evolution of the product itself. Over the years, operators in the youth accommodation space – whether hostels, PBSA or youth tourism – have made significant strides in adapting to their audiences. This isn't just about staying competitive in increasingly crowded markets; it's about truly listening to what young people want and delivering on those expectations.

A major driver of this evolution has been the rise of online information, peer reviews and social media. Platforms like TripAdvisor or Instagram, alongside review systems embedded in booking engines, have completely changed the game. Young people today have unprecedented access to information and they're incredibly vocal about their experiences. This transparency means it's become much harder for operators to hide behind a substandard product. If the quality isn't there, the feedback will surface quickly and it can significantly impact demand.

That said, this digital shift has also created opportunities. If you're paying attention to these channels and actively listening to your market, it's easier than ever to refine and improve your offering. Customers are constantly sharing what they like and dislike – whether it's about facilities, community experiences, or even smaller operational aspects. As an operator, you've got a constant stream of data at your fingertips and the key is to use that feedback to create a better, more tailored product.

The shift from a more transactional approach to one that prioritizes experience, community and quality has been one of the most noticeable changes. And that's true across both sectors, whether you're talking about hostels catering to backpackers or PBSA serving students.

Generational trends shaping youth travel

Stephen: Yeah, that's a great point. One of the recurring themes in this book is how young people's approach to travel has evolved. Compared to when you and I travelled 20 years ago, it's so much less accidental now. Back then, we might have turned up somewhere with a Lonely Planet guidebook in hand, only to find the place we'd picked didn't even exist anymore. Now, they know the experience beforehand, often down to the smallest details.

I imagine that one of the key differences between running a hostel and PBSA is the length of stay. In a hostel, the average guest might only stay two or three nights. But with PBSA, you've got students living in your properties for a year, maybe longer. That must bring a lot more pressure to make sure they're happy over a much longer period of time. Have you noticed a shift in what young people want, particularly in the PBSA sector, in terms of facilities, attitudes towards socializing and their overall expectations?

Neil: Definitely. And when we focus on PBSA specifically, the expectations and horizons of the residents are quite different. As you said, they're committing to a year, maybe two or three, as opposed to the more transient stays of a hostel or backpacker.

Because of this, the decision-making process is much more careful and deliberate. We find that students often make these decisions collectively, involving their parents, grandparents or even siblings. It's a significant investment – financially and emotionally – and it can have a big influence on their university experience.

There are so many considerations that go into choosing PBSA. For example, proximity to their university is a big factor. How far is the accommodation from campus? What are the transport links like? And then there's the question of how well the building supports their academic needs. They're asking about study spaces – are there small private rooms, larger communal ones, or both? What's the gym like?

This is where the shift becomes apparent. When we think about how you and I might have been back then versus students today, it's fascinating. Students still love events, but now the focus has shifted. It's less about parties and drinking culture and more about cultural experiences, food-related events and other activities that foster connection and learning. Health and wellbeing are much bigger priorities now. We get far more questions about gyms than we ever used to, for example.

Because of this, the design and provision of amenity spaces are critical. Students are making carefully considered decisions and they expect the facilities to reflect their priorities. It's not just about having a nice-looking property; it's about meeting their practical, emotional and academic needs.

This contrasts sharply with the motivations and decision-making process of a backpacker, who might only be staying for a few nights or weeks. PBSA students have a much longer-term focus and their needs reflect that. They're not just choosing a place to sleep – they're choosing a place to live, study and grow for a significant part of their lives. That's probably the biggest difference I've seen.

QUESTION

Does your product lend itself to the types of visitors to your destination? If not, what needs to change? Do you have a health and wellbeing strategy for your customers?

Stephen: Cool, and with Scape having such a large portfolio in Australia and also in the UK, where your focus is, I imagine you must see similarities in the types of nationalities staying at your properties across these countries. But do the demands differ significantly between the two? Australia and the UK are quite different culturally and geographically, so are students looking for different things in each location? Or is there a common thread, like a focus on health, wellbeing and cultural events?

Neil: That's a great question and the answer is a bit of both. There are certainly generational connections and global trends that you see across all markets, but there are also notable regional differences.

Let's start with the commonalities. As a generation, today's students are generally more health-focused, more committed to their studies and more altruistic in their concerns, particularly around sustainability and the environment. These priorities are consistent whether you're in Australia, the UK or the US. There's a shared consciousness among young people that influences what they value in their living spaces, from gym facilities to eco-friendly practices to spaces that support their academic goals.

That said, local differences absolutely come into play. Each region has its own demographic nuances and market preferences. While we do share markets to some extent – for example, students from specific countries feeding into both Australian and UK universities – you'll also find that different nationalities gravitate towards certain locations over others.

We design our products to be flexible enough to cater to a variety of demographic backgrounds and needs. Affordability is a big part of this. We don't want to create spaces that only cater to one small segment of the student population. Striking the right balance between quality, affordability and accessibility is central to what we do. It's about offering a broad range of options that resonate with students from different walks of life.

When you look at Scape properties, whether in the UK or Australia, there is definitely a core DNA that runs through all our buildings. The design, the amenities and the overall experience maintain a certain level of consistency, which is intentional. We want our product to be strong enough to deliver on those universal student needs, regardless of location.

At the same time, we make sure our offerings are adaptable to local markets. What resonates in Sydney might not have the same appeal in London or Leeds, so we adapt to meet those local needs while staying true to our brand values. It's a balance between consistency and flexibility, and I think that's key to serving such a diverse and global student population effectively.

The importance of long-term planning

Stephen: Neil, I think that's fascinating. You've had experience working in a large hostel chain like Generator and I've observed how, in Australia, there's a kind of ecosystem where travellers follow well-known routes – clockwise or counterclockwise around the country. It's almost predictable. Europe, on the other hand, is so different. With 30-odd countries and now the UK separated post-Brexit, there are countless paths people can take, which makes it harder to predict and capture specific nationalities year-on-year.

In contrast, when it comes to long-term education and student housing, it feels much more structured. Certain cities and universities consistently attract specific markets because of their academic specialties. From a marketing and operational perspective, I imagine the planning around student housing is much more defined than for hostels in Europe, which operate in a much broader and more unpredictable landscape.

I've also noticed that student housing has evolved dramatically since I lived in it between 2000 and 2003. Back then, it was... well, let's call it 'very social'. In my work with US students, I've seen a preference for shared dorms. And in Australia, when I lived there, my friends at the University of Sydney stayed in what felt like a very traditional, almost British boarding-school setup.

It seems like a lot has changed over the years, driven by both parents' expectations and the opening up of the global education market. Where do you see things going? What do you think students will want 8 to 10 years from now? And how might this impact both the educational experience and housing?

The role of feedback in innovation

Neil: That's the million-dollar question, isn't it? How do we predict the needs of students 8 to 10 years from now and create products that not only meet future expectations but also stand the test of time? Because, in our industry, the product lifecycle means you want your offering to be somewhat evergreen. You can't completely redesign buildings every time a trend shifts.

If I think about where student housing is now compared to when we were students, the evolution is stark. Take the University of Sydney's beautiful, historic halls – they're incredible spaces, but the needs of today's students are so different.

What we are seeing now is a blend of desires. Students crave community and social connection, but they also want privacy and high-quality facilities. When I think back to the halls of our generation, they often fell short in terms of quality and design. They didn't always meet the fundamental needs of students, at least not by today's standards.

What has changed is that students are far more discerning now. There used to be this assumption that students would accept whatever they were given because they did not have the purchasing power or options to demand more. That's no longer the case. The global nature of education and changes in students' purchasing power have completely shifted expectations.

This shift has been great for the sector because it has raised the bar across the board. Whether students are choosing premium accommodation or more affordable options, the overall standard has improved significantly over the past 20 years.

We also talk a lot about health and wellbeing in the sector, and that's a huge driver of change. The cold, uninviting student accommodation of the past isn't acceptable anymore – not for students and certainly not for parents. Today, the emphasis is on creating environments that promote wellbeing, whether that's through better lighting, better social spaces, or simply offering more modern and comfortable facilities.

Another key evolution has been diversifying the housing offering. Not every student wants the same experience. Some international students or postgraduates value a quieter, more private setup, such as a self-contained studio. Others, particularly undergraduates, want a vibrant social life. For this group, clusters – shared apartments with communal kitchens and living spaces – have become a cornerstone of modern student housing.

Clusters allow us to create micro-communities within a larger building. This is especially critical for first-year students who are living away from home for the first time. It's not just about giving them a roof over their heads; it's about creating an environment where they can make friends and feel part of a community, minimizing loneliness and isolation.

The provision of amenity spaces has also evolved significantly. It's no longer enough to have a basic common room. Students now expect thoughtfully designed spaces for studying, socializing and exercising. Gyms, communal kitchens and breakout areas have become standard in many buildings.

To keep up with these changing needs, we constantly engage with our students. We run focus groups, surveys and feedback sessions to understand what's working and what isn't. Their input directly influences our decisions – whether we're refurbishing an existing building or designing a new one.

So, looking ahead, I think the focus will remain on balancing community, privacy and quality. As new generations of students come through, their preferences will continue to evolve and it's up to us to stay ahead of the curve while keeping our products flexible enough to adapt.

Stephen: Yeah, and I guess this evolution, which we talked about 15 years ago, has been mirrored in hostels as well. They've had to evolve for many reasons. For instance, the 100-bed dorm in Fiji with no facilities might have been fine 20 years ago, but it's not necessarily what travellers want today. Flexibility and adapting the offering have become crucial to staying relevant in an ever-changing market.

One of the biggest challenges over the last 20 years for the sector – whether hostels or PBSA – was, of course, Covid. Both rely heavily on international guests, and when the world stopped moving it was devastating.

Have you noticed any trends from the Covid period that have persisted? And, on the flip side, were there any concerns that emerged during Covid – like the fear that people wouldn't travel anymore or that everything would shift online – which have since dissipated? It feels like the world has adjusted, and while some habits might have changed permanently others haven't stuck.

Neil: Yeah, Covid was a hugely difficult time for so many sectors, including ours. It's strange to look back now and realize how recent it was and how quickly it came – and then, relatively speaking, how quickly it went away. In some respects, things have returned to 'normal', or at least what we all perceive as normal. But, in other ways, Covid has left its mark on behaviours and expectations.

One area where we've seen changes is in market behaviour, particularly in booking patterns. For example, there's now a stronger desire for assurance around cancellation policies and flexibility – people want to know they can adapt their plans if needed. I think this shift in purchasing behaviour is one of the lasting impacts. Customers are looking for confidence and clarity when they book, and we've had to adapt to meet those expectations.

What hasn't changed – and this is reassuring – is the youthful desire to travel and explore. Despite all the fears during Covid, we haven't seen a decline in key markets' willingness to go abroad, study or experience new things. That spirit to travel, acquire new experiences and connect with different cultures hasn't been dampened. In fact, it seems as strong as ever.

Covid didn't fundamentally alter the desire for international experiences – it just temporarily paused it. What it did change is how we, as operators, approach the market. For example, we've become more proactive about offering flexible booking options and building trust with customers. People want to feel comfortable booking their plans and we've adapted to provide that assurance.

If there's a silver lining to Covid, it's that it pushed us to rethink how we serve our customers. By introducing more flexibility and better support during the booking process, we've not only met the demands of the post-Covid traveller but also strengthened our customer relationships. Hopefully, these long-term trends will continue to evolve positively.

Stephen: Cool. Well, that's interesting and it touches on one of the recurring themes throughout the book: the importance of creating an ecosystem – whether it's your brand or your destination – that works for young people. Flexibility in booking, for example, is just one part of it.

I guess one of the things that's come up time and again is how differently governments or destinations value the long-term impact of young people visiting their country. Across the US, Australia and the UK, have you noticed any differences in their openness – whether that's in terms of granting planning permission for PBSA or fostering youth tourism more broadly?

Neil: We definitely see differences, both in policy and in the overall attitude towards youth-focused products, and these vary over time. Government policies, particularly around visas or openness to study, tend to ebb and flow depending on the administration in power. These policies can have a direct impact on student numbers and the broader youth travel market.

As operators, we're constantly working to influence policy, trying to ensure that governments understand the long-term value of welcoming young people to their destinations.

The question of planning permission is particularly interesting, or perhaps better framed as perceptions around youth-focused products. Whether it's hostels, PBSA or other offerings, we have often faced the challenge of explaining to local stakeholders that these can be high-quality, well-run operations that bring real value to the local economy and community.

Take PBSA as an example. Over time, it has evolved into an institutional-grade product, which was not always the case. In the past, PBSA was seen as a more alternative, niche sector. In the UK, this perception has largely shifted and there's now a broader understanding of the benefits it brings. However, in parts of Europe, PBSA is still a nascent concept and there is a lot of work to be done to educate decision-makers about its potential.

Similarly, with youth tourism and hostels, I remember how, when we were opening hostels in major cities like Paris and Rome, we had to show local authorities what a modern hostel could look like. There were a lot of preconceived notions about what a hostel was – associations with low quality or a lack of safety – and it took effort to overcome those perceptions.

What is interesting is that the market for hostels has also evolved. Traditionally, hostels were the domain of backpackers. Now, they appeal to a much broader audience, including city breakers and people who might not have considered a hostel 20 years ago. Hostels have embraced this shift, offering a range of options – from social dorms to high-spec private rooms and pod-style accommodation.

At the same time, hostels have retained their core strengths: fostering a sense of community and offering unique social experiences. Most have done an excellent job of balancing affordability with high-quality amenities, which has helped shift public perception.

Both PBSA and hostels are unrecognizable from what they were a generation ago. The products have evolved so much and the quality, flexibility and diversity of what's on offer now are remarkable. I don't know if you have had a similar experience, but for me it has been incredible to see this transformation first-hand.

From students to professionals

Stephen: Having worked across different fields, I've noticed that when something is perceived as niche – like PBSA or hostels – it can be hard to change perceptions, even though the scale and economic impact of these sectors are anything but niche. When you look at the revenue generated for countries and local communities or the sheer scale of operations, it's massive.

You're not managing a couple of HMOs in London; you're looking after thousands of young people. And, as more institutional investment has poured into hostels and PBSA, it's clear that the capital markets have become more open to understanding what these products represent. That openness even extends to things like planning permission, because without sufficient student housing universities struggle to grow, shops lose potential customers and local transport networks don't get as much use.

There are some cities that physically can't accommodate more students because the infrastructure isn't there. Sometimes that's because local governments or financial systems don't understand the opportunity, or one part of the ecosystem moves ahead while the other lags behind. It's been well-documented across Europe that the post-Covid boom in students returning to their studies has led to unfortunate scenarios where international students haven't been able to find housing near their campuses. The knock-on effects of not having a functioning ecosystem are significant.

What's your take on how this has changed? I mean, we both worked at Wake Up! Hostel in Sydney, which was trailblazing for its time. It had public spaces, large dorms and scale – 600 beds when it launched in 2002. It's amazing to see how much the product has matured since then. Hostels and PBSA have evolved from niche markets into polished, mass-market offerings.

One thing I'm curious about is Scape's move into co-living with Morro. Was that driven by the idea of building long-term brand loyalty, similar to how airline alliances aim to get you at the back of the plane with the hope you'll move to the front over time? Or is it more about meeting a growing demand for flexible housing options in cities?

Neil: That's a great question and I think the drivers are a mix of both. On one hand, there's a natural lifecycle to our core student audience. Once their studies are complete, it often doesn't make sense for them to remain in student-specific housing – for planning or other reasons. But beyond that, we've recognized that the fundamentals of the PBSA product – high-quality amenities, a sense of community and flexible living – have a much broader appeal.

There's a growing demographic of people who value those same attributes, even if they're not students. They're looking for flexibility, adaptability and dynamic living environments that meet their needs at different life stages.

So, we've taken the DNA of what works in PBSA – designing for community, great amenities and a focus on wellbeing – and pivoted it towards a more general audience. It's not about watering down the student experience; it's about building a product that resonates with a wider cross-section of people.

We launched our first Morro site in 2021 and the response has been fantastic. We've got a strong pipeline in London, the US and beyond. It's still early days for co-living, particularly in the UK, but it reminds me of where we were 10 years ago with PBSA. Back then, we had to educate planning offices and local councils about what student housing could look like. Now, we're having similar conversations about co-living.

I remember taking planning officials from one London borough to our Guildford site a few years ago. Once they saw the product in person, they were completely sold. It only took that one visit to get them on board. Sometimes it's just about getting people to see and experience the product for themselves.

As for the long-term brand loyalty angle, that's definitely part of it. If we can create a relationship with someone during their student years and continue to support them as their needs evolve – whether through co-living or other products – that's a huge win. But, at its core, it's also about recognizing that the demand for flexible, high-quality housing is only going to grow and we want to be at the forefront of meeting that need.

TIP

If you are struggling with getting local government to understand your new plans, why not invite them to go and see what the product is like in real life. Youth accommodation products are often misunderstood but can be important for the youth travel ecosystem in a destination.

Stephen: It's such a broad spectrum, from school trips for 11-year-olds to working holidaymakers in their early thirties. What ties it together is the independence and exploration, even if the level of independence varies with age or life stage.

The flexibility you mentioned is absolutely critical. Gone are the days of rigid models where accommodation only catered to one specific type of traveller or student. The modern approach, whether it's in hostels, PBSA or co-living, is about offering options. That could mean a 12-bed dorm for budget-conscious travellers alongside private suites for those looking for more comfort; or creating shared spaces that meet the social and practical needs of students and young professionals alike.

It's also a reflection of how much the expectations of young people have evolved. Today's generation has grown up with choice. They're used to products and services that adapt to their needs, not the other way around. So, whether we're designing student housing or co-living spaces, that adaptability has to be built into the model.

The focus groups and feedback loops we use are critical to this process. It's not just about predicting what people might want in the future; it's about staying constantly engaged with their changing preferences and behaviours. And it's been fascinating to see how those preferences differ slightly across markets but still reflect some universal generational themes – like a focus on wellbeing, community and flexibility.

I've really enjoyed this conversation, too. It's always great to take a step back and reflect on the bigger picture, especially with someone who's been in the industry for as long as you have. These are important conversations to have because they push us to think about how we can evolve and improve.

Conclusion

The conversation with Neil highlights a pivotal truth: youth-focused accommodation has evolved from niche offerings into highly strategic, diverse products that address the needs of a generation defined by choice, flexibility and global aspirations. Whether accommodating backpackers in hostels, students in PBSA or young professionals in co-living spaces, operators are embracing change and prioritizing wellbeing, adaptability and community. The Covid-19 pandemic was a moment of reckoning, accelerating some shifts while cementing the enduring demand for meaningful experiences and global exploration. As the sector matures, collaboration among operators, governments and communities will be critical to creating ecosystems that support the next generation of global citizens.

KEY TAKEAWAYS

1 **Evolving expectations of young people:** Today's youth prioritize flexibility, health, wellbeing and community in their accommodation. The modern traveller is discerning, with expectations shaped by access to reviews, social media and peer recommendations.

2 **The longevity of youth travel demand:** Despite fears during Covid, the desire to travel, study and explore internationally remains strong. Booking patterns now reflect an increased demand for flexibility and assurance, underscoring the need for trust and adaptability in the sector.

3 **The rise of co-living as a flexible housing option:** Co-living bridges the gap between student housing and professional living, catering to young professionals seeking community and convenience. Operators like Scape leverage their PBSA expertise to create adaptable, high-quality products for diverse audiences.

4 **Government and ecosystem challenges:** Youth-focused products like hostels and PBSA face hurdles in gaining recognition and support from local governments and planners. Destinations must align their infrastructure and policy to attract and accommodate international students effectively.

5 **Feedback as a driver for innovation:** Constant engagement with students and residents through surveys and focus groups drives product innovation. The shift from rigid accommodation models to flexible, multi-faceted spaces reflects the sector's responsiveness to changing needs.

6 **The global connection:** While generational trends like sustainability, wellbeing and community are universal, local nuances influence product design and preferences. Successful operators balance global consistency with regional adaptability to serve diverse markets effectively.

7 **A maturing sector:** Hostels, PBSA and co-living have evolved into polished, institutional-grade offerings with mass-market appeal. These products are increasingly recognized as critical components of local economies, supporting universities, tourism and community growth.

14

The shifting landscape of youth marketing

This chapter delves into an in-depth conversation with Simon Lucey, a UK-based marketing expert with extensive experience in understanding and reaching the youth demographic. We explore the shifting landscape of youth marketing, emphasizing why understanding the subtle nuances between age groups and how they interact with technology is paramount for a brand's success. We unpack lessons learnt from working with major brands, reveal the power of young people as influencers within their families and communities, and examine how long-term investments in this demographic can yield significant returns.

Stephen: Can you give us some background in terms of your career in youth marketing and what you have been up to.

Simon: My career in youth marketing really began when I was a student. I worked for a ticketing company, selling tickets for club nights. During that time, we were approached by BlackBerry with an incredibly challenging brief. They wanted us to convince everyone that touchscreen phones were just a fad, something that wouldn't last. So, we were out on campus spreading the message that BlackBerry was here to stay. In hindsight, I think it's safe to say we weren't very successful in that first mission!

From there, I went on to establish a company called Hype Collective, a youth and student marketing agency. We worked with a mix of travel and consumer brands, as well as employers looking to market to 16- to 25-year-olds. We worked with brands like Co-op, Disney and Adobe. That company was acquired last year and I've just recently finished my time there. Now, I'm exploring my next steps after spending the last seven or eight years figuring out the best ways to communicate with that 16- to 25-year-old demographic.

Redefining youth: The challenges of broad age brackets

Stephen: When discussing how to quantify the 'youth' demographic, it's interesting that you refer to the 16- to 25-year-old age range. While researching for this book, I have found that there is no universally agreed-upon definition for youth travel. As a result, I have adopted a broader range, defining youth as those between 11 and 35 years old. This spectrum includes school groups starting as young as 11, who might go on school trips – semi-independent in that they are independent of their parents, but still under the supervision of teachers. The range extends up to 35 years old due to updates in Working Holiday visa regulations.

You may be too young to remember, but there used to be brands such as 2wentys or Club 18–30 where, once you turned 31, you were no longer eligible. Although those brands are no longer around, they would have likely evolved into what we might now call 18–35 holidays.

When it comes to marketing, particularly on the digital side, understanding these age groups is crucial. There are different challenges depending on which specific segment within the youth demographic you are targeting. In fact, 16–30 might now be considered too broad, given how much technology has shaped behaviour. A 16-year-old's engagement with marketing is often vastly different from that of a 30-year-old.

TIP

This shows the breadth of knowledge needed to really create powerful marketing campaigns. Although I defined youth travel quite broadly in earlier chapters, there are niches within this sector that require particularly skilled marketing strategies.

Simon: Absolutely. I have always operated with a relatively narrow focus, but I recognize that there are valuable insights to be gained from a range of ages, from 12-year-olds to 30-year-olds. That said, I think relying solely on age as a demographic marker in marketing is somewhat lazy. For instance, I was surprised recently to find that my dad has been ordering items from TikTok Shop, while my much younger nephew has never made a purchase there. Age-based stereotypes don't always hold true.

There is certainly a marketing challenge in how we segment these audiences – where we likely need to be more precise about who we're targeting. However, when it comes to destinations and government policies, the benefits of attracting more young people to a country might be better served by adopting a broader, more holistic perspective.

I think that goes beyond travel as well. Everyone has a different definition of what 'young' is, and it depends on many different factors. People will argue about it all day and, in all honesty, if you are happy with it that is all that matters.

Lessons from the field: Key success factors in campaigns

Stephen: You have worked with many travel brands and destinations on campaigns. What were the most successful ones that you worked on, and why do you think they were a success?

Simon: As I was walking here for this interview, I found myself reflecting on the patterns that have emerged among successful campaigns, particularly from the client's perspective. What stood out was that successful clients all had one senior person who truly understood the core problem or opportunity. This person could distil that understanding down to their team and to the agency.

Sometimes, the problem or opportunity was straightforward. For example, we worked on a campaign with the Northern Territories where they told us, 'We've spent all this time building a brilliant nurture pathway and we're confident that once we get email sign-ups, we'll see action from it.' Their brief was simple: they needed more email sign-ups. That clarity made it easy for us to focus on the task.

On the other hand, we have had clients who were certain that their problem was related to a specific channel, like Google. For instance, when we worked with StudentUniverse, their Google acquisition costs were rising and they needed us to explore other channels where people were searching, so we could position the brand effectively. They were clear about the problem, even if they hadn't fully identified the opportunity yet. This clarity allowed everyone to pull in the same direction.

From an agency perspective, our role is to solve problems. While there is always competition between agencies to see who can do it best, the key to success is grasping the true nature of the problem. That is a challenging skill, especially in today's complex world with countless channels and data points. Senior leaders who can cut through the noise and bring simplicity to a complex situation are invaluable.

The biggest challenges come when internal stakeholders have different understandings of what the problems are. In that situation it's our job to try to identify the problem and convince everybody internally that we've identified the correct problem!

TIP

Find a channel that works best for you. It may require a number of different channels with different strategies for each but keep an open mind and keep nimble.

Youth as influencers: Beyond direct ROI

Stephen: Without having those ambassadors within the destinations or brands who truly believe in the value of the youth market, it can be challenging to fully unleash your creativity or deliver what you know you are capable of. When it is approached with a mindset of just ticking a box, the youth market is particularly adept at seeing through that kind of attitude when it comes to campaigns.

Simon: Yes, I mean, fundamentally the work is not as good. We have had briefs in the past where we have worked with somebody who potentially has not got the brief or buy-in from the senior contact and then there is a great challenge to get the best out of a campaign. There has clearly been a bit more of an apathy from client side throughout the entire process and the result is always much worse. I think having that senior buy-in is so important. This is something I know that you speak about all the time – raising the profile of the youth market to a senior leadership level. It is crucial to good work being done, whether it is inside marketing departments or in the product or delivery, or any aspect of the business.

Striking the balance: Short-term wins vs long-term gains

Stephen: One of the themes of the book is advocating for brands and destinations to adopt a long-term perspective when investing in youth travel. It's not just about securing email sign-ups; if a destination is great but unknown and visitors have an amazing experience, they'll want to return. Have you encountered any briefs where destinations or brands were thinking more long-term? Were they aware they were running short-term campaigns primarily to get young people engaged initially, but with a broader strategy in mind? Did you have those kinds of conversations with clients? If so, could you share any examples?

Simon: I often found it challenging to initiate conversations about investing in young people now with the expectation that they will spend more later. For instance, outside of the travel industry, banks historically invested large budgets to acquire student banking customers. They offered giveaways and freebies, especially during fresher's week at universities, with the hope that these students would later purchase mortgages and build their savings with the bank. However, in today's world, pitching this idea to a financial director can be difficult. If you suggest that this strategy will impact the bottom line in five years, you might not be taken seriously.

However, there's another angle that's an easier sell to finance departments: young people as tastemakers and influencers within their families. The power of what happens on platforms like WhatsApp can be substantial for brands. I read an analysis of the UK general election, which pointed out that while Labour's TikTok account didn't significantly shape young people's voting habits, it was effective in other ways. Many young people shared Labour's TikTok memes within their family WhatsApp groups, sparking conversations that influenced the voting decisions of older family members. This principle applies to young people driving demand within their families as well.

There's certainly opportunity here. Take, for example, companies like Hopper, a travel app. They sell relatively low-margin flights but have built additional products and services around their core offering. I believe most of their revenue now comes from these financial add-ons rather than from selling low-margin flights. So, there are opportunities to build revenue once you've acquired that initial customer. The key is that this customer needs to generate enough revenue in the first year, but there's also a long-term benefit to acquiring young people early, which can make your business more sustainable over 5 to 10 years.

Stephen: In another interview, we discussed the aviation industry and how getting a young person on a plane with the cheapest fare can potentially foster loyalty throughout their lifetime. Ideally, they'll eventually pay for tickets further 'up the plane' for business or leisure travel.

Having worked across multiple industries, I assume that everyone you work with is focused on their return on investment (ROI). There's the direct ROI (e.g. when you track a purchase that comes as a direct result of a campaign) and then there's the indirect ROI, which includes referrals, third-party endorsements, or young people who have a great experience and then become influencers – essentially doing the marketing for you without additional cost. I assume that kind of indirect brand sharing was crucial to the campaigns you were running.

Simon: Yes, and you can see it in the way that a lot of Middle East destinations are marketing themselves with that heavy investment in sport. Clearly, sport isn't just for young people, but it commands a large youth audience. They are clearly being successful in using that as a way of growing a young audience. The stereotype is obviously that younger people spend less per head when they're out there, but while they are acquiring that young audience, they are also driving those that are older as well.

Adapting to change: Learning from sectors outside travel

Stephen: I'm older than 36, but I've been working in hospitality and travel for over 20 years now, mainly within hotels. I've noticed that they often lag behind when it comes to marketing segmentation and adopting new tools like social media. You've worked outside of travel – are there particular sectors that do this better? If so, could you share some good examples?

Simon: I kind of agree, but I can't quite pinpoint why travel and hospitality don't seem as innovative. When I think about industries that are really pushing boundaries, it's often those that operate under real constraints that must be the most creative.

Take, for example, the charity sector. We might stereotype it as lacking big budgets, but that very limitation forces them to be innovative. One standout organization is the Campaign Against Living Miserably (CALM). With a tiny team and almost no budget, they manage to run massive national awareness campaigns every couple of years, far exceeding expectations. I can't think of any travel organization that could deliver such a powerful campaign on a similarly tight budget.

On a different end of the spectrum, consider Lovehoney, the sex toy website. They face their own constraints because they can't advertise on major platforms like Meta or TikTok. This restriction has forced them to innovate far ahead of the curve in influencer marketing. When we started working with them four years ago, we thought we were experts in influencer marketing, but their internal team was ahead of us because they had been compelled to explore alternative channels.

So, perhaps the travel industry's problem is that it's almost too easy. Holidays naturally look great, and shooting a video of a holiday is relatively straightforward. Industries that face barriers are compelled to innovate.

TIP

You do not always need big budgets to create amazing campaigns aimed at a younger audience, but you do need buy-in from top to bottom of the organization to make it work and believe in that strategy.

Stephen: So, I think what you're suggesting is that those organizations working under tougher constraints are often forced to think outside the box and come up with better solutions. This aligns with another interview I had with Emma English from BETA, in Chapter 7, where we discussed how the UK seems to have always faced barriers – whether it's the lack of government support for youth marketing, the high cost of visas, Brexit or other challenges.

I suppose the operators, organizations and businesses within youth travel in the UK have had to find ways to get their brands out there and attract young people, despite these obstacles. Perhaps these restrictions do inspire creativity at times.

Simon: I certainly don't think we should wish too much adversity on the youth travel industry, and I'm not suggesting that this is a justification for government policies that could push us backwards. However, the industry is remarkably resilient and people often come up with smart solutions when faced with challenges. Without those challenges, we might just opt for the easy, obvious answer, which isn't necessarily the most creative one.

Stephen: If you were advising a brand or destination that's about to launch a marketing campaign for the first time – or a brand that's realized they need to lower their average customer age, which is quite common in hotels – what would your advice be? How can they avoid overcomplicating things, and what are some pitfalls they should steer clear of, especially if they're targeting youth for the first time?

Simon: When you're embarking on a new marketing campaign, especially if you're trying to reach a younger audience for the first time, it can feel incredibly overwhelming. There are countless options available, and it's easy to get bogged down in trying to do too much at once. Regardless of the size of your organization, you'll often feel under-resourced. I recently spoke to someone at a large sports brand who said that they don't have the budget compared to their biggest competitor. I remarked, compared to 99.9 per cent of other marketeers, they're doing pretty well! The reality is that almost every organization, whether it's a one-person operation or a global brand, feels like they don't have enough resources.

So, the key is to be strategic about where you invest your time and resources. I recommend taking a very methodical approach. Start by listing out all the possible channels you could target on a whiteboard. Then, rank them based on two criteria: your confidence in your ability to execute a successful campaign in that channel and the potential results that a successful campaign in that channel could yield.

You'll likely find that some channels are familiar and you're confident you could run a strong campaign, but the potential scale might be limited. For example, you might have a great pay-per-click (PPC) expert on your team and feel confident about running a PPC campaign but realize that there isn't enough search volume to drive the results you need. On the other hand, you might identify channels with massive potential reach, but you're less confident in your ability to execute well because you haven't used them before.

Where these two factors overlap – where you have both confidence in your prediction and potential for significant results – that's where you should double down. It's essential to focus on just two or three core channels, even though it might feel counterintuitive. Marketing today is full of distractions; you might see a competitor launching a massive Snapchat campaign and feel the urge to jump on that bandwagon. But if Snapchat isn't one of your chosen channels, that's okay. It's better to maintain focus.

At the same time, it's wise to run smaller tests in the background with other channels to ensure you're on the right track. But, ultimately, having a singular focus on a smaller number of channels than you might initially think necessary is crucial. Identify the two or three core channels that are most important for your client recruitment, and invest in training and development around those.

You might also have a few secondary channels where your goal is simply to be satisfactory. These are channels where you're not aiming for big wins and it's okay if the results are just okay. By separating out these secondary channels, you avoid the trap of over-investing time and money in areas that won't deliver the same return on investment.

The technological divide: Navigating platform preferences

Stephen: Do you think a multi-channel marketing approach is more essential when targeting youth, or does it depend on the specific age group or demographic you're aiming for? For instance, you might assume your target audience is primarily on TikTok, but they could still be more active on Instagram, especially in countries where TikTok is less popular or restricted by government regulations. How important is it to consider these factors? Do you typically recommend brands conduct thorough research before deciding on platforms, or do you advise them to follow competitors' presence on platforms like TikTok as a guideline?

Simon: There are still millions of people on Instagram – it's not like the app has disappeared. I'm currently working on a project involving Pinterest, which hardly ever gets mentioned. So, who's on Pinterest? A huge number of people – predominantly for things like fashion, interior design and cooking. It's huge for those areas, but I genuinely believe there's an untapped opportunity for travel brands. I can't think of many travel brands active on Pinterest and yet there's this massive audience there. I think we're often too obsessed with doing what everyone else is doing.

TikTok, for example, is clearly a huge opportunity. It's revolutionized all social media channels – just look at how Instagram now mimics TikTok's style. The key to success on TikTok is ensuring you've got a solid strategy. But there are also these underused platforms, like Pinterest, where others don't venture because they overlook it and head straight for the big players. Yes, TikTok is massive and full of opportunities, but even if Pinterest or Instagram has a fraction of the following, that would still be more than enough for most brands in terms of audience size.

TIP

Do you have a TikTok strategy for your brand or destination? Invest time in understanding the app and do not ignore it.

Stephen: I find it quite interesting how different people use technology, even within families. For example, my wife, who's Swedish, and her family use Facebook Messenger for their group chats, while I use WhatsApp with my sister and dad. Then there's my 19-year-old nephew – he'll hate me for saying this – but if I WhatsApp him, it can take a week for him to respond! On the other hand, when messaging friends or people closer to my age, even as young as 36 like you, we're all pretty quick to reply.

It really highlights how different demographics, age groups and cultures interact with technology and platforms. Understanding these differences is crucial before investing thousands of pounds or dollars into a marketing campaign. You can't assume everyone engages with the same platforms in the same way.

Simon: I think we assume that social media has got rid of fragmentation in many ways. Fragmentation in marketing means that brands can no longer rely on a 'one-size-fits-all' approach but must diversify and customize their outreach to resonate across various touchpoints where their audiences spend time. You know, before social media, it would be unthinkable to think that there's just one location that you go to with regards to marketing to young people. There are now just a few main platforms that really work for attracting young people to your brand. I was recently in the US with my cousins and said, 'I'll WhatsApp you,' and they said, 'I'll just redownload it again from the last time I was in the UK.'

Stephen: So, what were they using, out of interest?

Simon: Telegram! I really wasn't expecting that. I am not sure why they use that but think it might have been a reaction to Meta's policies and privacy concerns.

Stephen: That is something that I guess changed in terms of how intrusive brands can be, and young people are probably a little bit more sensitive than older generations with regards to their privacy and data sharing.

Simon: I think there's been a noticeable shift in trends over the past five years. It once seemed like everything was converging towards Meta and its suite of brands, with Meta and Alphabet essentially dominating the digital landscape. However, much to Mark Zuckerberg's disappointment, the last five years haven't unfolded quite as either of these tech giants had hoped.

Google, for instance, is feeling pressure from both sides – on the one hand from AI advancements, and on the other from TikTok and other emerging apps. This has resulted in more choice for users, and people naturally gravitate towards social platforms where they can connect with like-minded communities. Social media, in that sense, mirrors human behaviour throughout history – people move in groups and these digital shifts often occur on a large scale, with entire geographies or territories adopting a platform together.

Stephen: I guess it is human nature to be part of a tribe, and this shows that, even digitally, humans still behave in ways we have always done.

Whether it's travel or marketing generally, what do you think the next three years will look like for brands and destinations aiming at creating, growing or retaining a youth audience? Also, are there new tools or platforms on the horizon that people are talking about but not really using at the moment?

Simon: I believe TikTok's search functionality will soon become as integral to marketing as Google has been for years. As TikTok continues to evolve, we can expect to see businesses bidding on search terms like 'cheap flights to New York' on the platform. Both organic and paid videos will be utilized, and TikTok will likely improve its data insights, which are currently somewhat rudimentary. The release of more robust business data will significantly enhance our ability to track marketing campaigns effectively.

We haven't yet touched on AI, but I foresee teams focusing on how to integrate their content with AI technologies, particularly for destinations. For example, I recently used ChatGPT to plan a trip to Mexico. After spending hours researching various channels, I simply asked ChatGPT, 'My wife and I are looking for a two-week trip around Mexico, with at least five days on a beach, prioritizing good food and culture.' The result was strikingly similar to the detailed itinerary I had painstakingly created. This suggests that AI will play a crucial role in generating travel ideas and influencing search results.

For destinations, there will likely be dedicated teams working to ensure their locations feature prominently in AI-generated content. However, for individual travellers considering whether to go backpacking in New Zealand or Australia, these destinations might not have as much control over the content that appears in AI search results.

Evolving traveller segments: Tech-driven vs tech-free explorers

Stephen: Travel has changed significantly from the days when we used to travel with a guidebook and a lucky coin, with an element of uncertainty. Back then, the unpredictability of the journey – like arriving at a destination without a reservation and finding a place to stay on the spot – was part of the thrill. The anticipation and dreaming stage were integral to the overall travel experience. Today, much of that spontaneity is lost as most travel details are known before arrival, taking away some of the excitement that I still enjoy.

This shift poses a challenge for brands and destinations: how to maintain that sense of discovery and excitement for travellers who now arrive fully informed. In the past, we would visit restaurants based on recommendations, not knowing what would be on the menu. Now, many high-end restaurants refrain from posting daily menus online to retain some element of surprise, as too much information could deter potential visitors.

Given this trend, do you think storytelling will become increasingly important for brands and destinations to recreate that sense of adventure and intrigue?

Simon: I think it's super interesting. Hostelworld have done some quite interesting work around this. They have taken quite a social approach to how you book now, where you can actually see who will be in the dorm room. This is interesting because you'll be able to see if there's a big group of travellers that you think you'll get on with before you arrive. You're able to interact with them before, so actually you are cultivating the experience earlier because you're allowed to engage with the existing customers. I think there will be tech powered solutions that give us digital experiences that can enhance the physical.

But then the challenge is – where's the adventure when everything's cultivated? And it would be interesting to see, is that just something that

I care about because I had the adventure and I had that feeling of just rocking up in a town and being like 'Oh God, this hostel is closed. I'm going to have to trek across town'? That was quite fun, and is that something that will be lost with our generation.

Stephen: I still, to this day, love doing the research for trips or even finding a new place to eat, and you probably do, given your travel experience and age. You probably really enjoy the research but feel, wow, ChatGPT has almost nailed this and it was easier with a few tweaks. It's a choice for the person. Do they want that research, or do they want someone else to do it for them? And that is where back in the day is where you would go to a travel agent to book your whole trip or get a round-the-world ticket and hope for the best.

That is something for travel brands to consider as a high priority going forwards. I think one of the challenges that travel companies have now is that technology is changing so quickly. It also means trends are changing so quickly, which means if you don't have a flexible marketing and digital strategy you could go all-in on one focus, and by the time it's live an algorithm has changed and that campaign is redundant.

This ever-changing environment poses significant challenges not only for agencies advising brands but also for the brands and destinations themselves. There may even be a shift back to more traditional forms of travel, much like what we are witnessing in the music industry. As digital platforms like Spotify and iTunes led people away from physical music purchases, the resurgence of vinyl records among true music enthusiasts suggests a growing nostalgia and appreciation for tangible experiences. Similarly, this could signal a trend in travel where a return to more personalized and less predictable experiences could gain traction once again.

Simon: For many travellers, the way they book trips has evolved rather than fundamentally changed. Most people who once relied on travel agents or booked packaged holidays still do so, but now they utilize different technologies to facilitate the booking process. However, there is a growing niche of travellers who value the art of preparing their journeys without relying heavily on technology or external support. This group takes pride in building their travel experiences without constant digital assistance, choosing to explore with minimal technological interference.

While they might still use tools like Google Translate, they intentionally limit their tech use, appreciating the authenticity and unpredictability that comes with a less connected approach. This trend suggests a divide in traveller behaviour: the majority embracing tech-driven convenience and a minority seeking a more traditional, hands-on experience that celebrates the journey itself, free from the over-planning often enabled by modern technology.

Stephen: It will be fascinating to observe how rapidly and profoundly AI transforms the way we travel, particularly in managing aspects of life administration like travel planning. For some, including myself, there will always be joy in personally researching and discovering hidden gems – especially when visiting a new city or town for work. Finding that off-the-beaten-path restaurant still offers a sense of adventure that technology can't fully replicate for me.

Reflecting on the spirit of the early explorers of the 12th and 13th centuries, who ventured into the unknown without maps or expectations, there is a fundamental human desire for discovery that technology may never completely satisfy. As AI continues to shape our experiences, there may come a point when we feel that technology's influence has become too intrusive, prompting a shift back towards a more traditional, exploratory approach to travel – one that embraces uncertainty and the thrill of the unexpected. A bit like how young people now are less comfortable about sharing data with the big technology firms.

Simon: This evolving landscape presents a significant challenge for travel marketers, who must navigate an increasingly segmented market. On the one side are travellers who fully embrace technology throughout the travel purchasing and experience cycle, seeking convenience, personalization and instant access to information. On the other side are those who deliberately limit their interaction with technology, craving a more traditional and authentic travel experience.

Conclusion

In this chapter we delved into the evolving landscape of youth marketing and the intricacies of engaging younger audiences in an era of constant technological change. The insights provided by Simon Lucey underscored the necessity for brands to not only be agile and innovative but also deeply understand the nuances of their target demographic. Brands that can navigate this balance, with a clear strategic focus and an openness to embracing change, will cultivate long-term loyalty and sustain their relevance in a rapidly shifting marketplace.

KEY TAKEAWAYS

1 **Defining the youth demographic in travel:** This conversation highlighted the lack of a universal definition of 'youth' in the context of travel marketing. While Simon focuses on the 16- to 25-year-old range I consider it a broader spectrum of 11–35 years old, reflecting the complexities of marketing to different age groups within the youth segment. Simon emphasizes that age alone is too simplistic a target market, with consumer behaviour often defying age-based stereotypes, necessitating more niched audience segmentation.

2 **Key success factors in campaigns:** Simon identifies that successful travel campaigns often have a clear understanding of the core problem or opportunity, guided by a senior leader who effectively communicates this to their team and the agency. He cites examples such as the Northern Territories' straightforward focus on generating email sign-ups, demonstrating how clarity of purpose allows for effective execution. He also talks about how campaigns with vague briefs or lack of buy-in from senior leadership often result in less impactful work.

3 **Challenges and opportunities in youth marketing:** A recurring theme is the tension between short-term goals and long-term investments in the youth market. While some brands and destinations understand the value of engaging young travellers early, many struggle to justify the ROI in the immediate term. Simon highlights the importance of seeing young people as influencers within their families and communities, capable of driving demand and shaping broader consumer behaviours.

4 **The impact of technology and AI on travel:** You can see how technology, particularly AI, is reshaping travel planning and experiences and the balance between the convenience of digital tools and spontaneity in travel. However, there is a growing niche of travellers who resist over-reliance on technology, seeking authentic experiences reminiscent of early exploration. For marketers, this evolving landscape demands adaptable strategies that cater to both tech-driven and tech-averse segments.

5 **The future of youth marketing in travel:** Platforms like TikTok will continue to disrupt traditional marketing channels, with search functionality becoming as critical as Google's has been. The integration of AI in content creation and search results will further influence travel decisions, prompting destinations to focus on visibility in AI-driven recommendations. Marketers will need to balance the allure of new technologies with the enduring appeal of adventure, discovery and authentic experiences.

15

Engaging the next generation

Building loyalty and legacy through youth-centric tourism

In this chapter, I converse with Tom Cassidy, the Director of Tourism for Liverpool Football Club, about the importance of attracting youth travellers, even for a brand as renowned as Liverpool FC. Football, along with other sports, media and entertainment, serves as a significant draw for young people across various age groups. Tom highlights not only the crucial role of youth in the travel industry but also emphasizes the necessity of continuously evolving both marketing strategies and product offerings to avoid losing relevance in appealing to this demographic.

Stephen: Can you just explain your role at Liverpool FC and what you get up to?

Tom: My role as Director of Tourism at Liverpool FC is quite niche within a football club and has been to devise, build and execute a full tourism strategy of the brand. My main role revolves around revenue generation, which is important for any business. We must keep maximizing what we are doing and devise new revenue streams. We do this internally, but we also work very closely with the city region and other different bodies such as Visit Britain. So, my job is the main interface with them and to connect with others on local, national and international markets.

The role of tourism in a global football brand

Stephen: Why does a club or a brand as big as Liverpool Football Club need a focus on tourism per se? I guess many would think, why have someone in this role as people will just come anyway.

Tom: Firstly, it's my sole role, so I am really glad that they have this focus on it otherwise I would not be with you here today! I mentioned revenue before, and I am not going to hide away from that. Last year, the last financial year to be more precise (which finished at the end of May 2024), we grew revenue. So just the tourism side of what we do, as a stand-alone business, we generated £7.4 million of revenue and hosted 400,000 visitors, which is a record for us, and, I think, for all other football stadiums in the UK. This means that as a stand-alone business within the club it stands on its own two feet. All revenue goes back into the club, and that's important. While that might only be a drop in the ocean in terms of the club's wider revenue, it all goes back into the current rules around financial fair play and other financial measurements. So, everything that we make on tourism is reinvested back into other parts of the club. We want to make sure that whatever we do, we do it in the best way that we can (a bit like the football team). We have a really big team of full-time staff working solely on tourism and lots of casual workers and support staff, too. This allows us to deliver all the tourism experiences and services that we currently have on offer.

It has been good for us that we've grown the business to where we are now and are employing people in tourism-related roles, meaning there is more revenue going back into the local economy. That is important for us as a club, as the club is really the heart of this local community.

If you look back around 25 years ago there wasn't a tourism offering that was as good as it is today. It was more of a hobby for the club, rather than a full-time business focus as it is today. When I started, I could see some easy improvements in terms of the tourism business focus to improve the fan experience.

Liverpool is a huge global brand and football club with an enormous fan base spanning the whole globe. Next month, I will be in Australia and Southeast Asia with Visit Britain, meeting agents and operators from that region. Our tourism offering has a global audience that we need to continue to share and update.

For the global visitor to Liverpool as a city, or specifically the club, we need to make it accessible if they come to visit. We are a football club first and we revolve around football, and that is our primary offer, but not everyone can get tickets for matches. So, this could be the only chance that they get to ever visit Anfield; we need to make it memorable for them regardless of when it is. Even if it's a Tuesday night in November, when the rain is coming in sideways and it's freezing and you know somebody is coming from the other side of the globe; we have to do everything we can, to make it memorable.

QUESTION

Even a brand as large as Liverpool FC has to constantly look at its product and tweak it for the ever-changing youth market. When was the last time you reviewed your offering and the feedback from younger visitors?

Engaging the youth demographic

Stephen: Football has changed massively since the 1980s and early 90s in the UK. Football was not always a nice place for young people and females to go to and experience. Now the football offering is generally a nicer, more family-filled environment during a match day, but I guess tourism and the infrastructure is important to make sure the whole experience is more memorable. How much emphasis for you and your role is there on young people and attracting young people to the tourism offering that Liverpool Football Club has?

Tom: I think that falls within the mix for us. We have a real range of demographics that come and visit us. When you say young people, we are slightly out of the city centre and so a lot of visits are done either with a group, with family or with friends. I think it's important for us to appeal to everybody.

In terms of younger youth travellers, they are often the decision-makers of the family. Younger visitors are really important; we have developed our tourism offer to make sure we both attract these demographics and also make sure they are engaged when they visit us. We now have much more diversity in the tourism products which appeal to everybody at all ages. The most mass-purchased item is the stadium tour, and the way that we have embraced technology on there has made the whole experience much more interactive. Rather than a static walk around the stadium, we have gamified the whole experience to make it far more engaging to visitors of all ages and demographics.

Many of the visitors doing the tour are under 30, and we must make sure it is relevant to them. The development of digital technology has impacted how the youth of today absorb information. So, what we offer needs to reflect this and it needs to be dynamic and have a bit of excitement.

The 'rock stars' of today, in my view and from what I see, are the footballers. Football fans want to engage with them as much as possible. So, we need to give our visitors, especially younger ones, what they want.

Apart from groups coming with families for a leisure reason, we also host school groups and university groups. Therefore, we had to develop a series of more in-depth educational tools that are focused on these sets of visitors. We have been working for quite a while on these and look to launch them on a wide scale soon, as we recently built a new museum in the stadium.

On our educational tools, learning through football and Liverpool Football Club, the research we have done has suggested that people of different ages learn more when it is linked into something that they have a real passion for.

TIP

Educational tools have to be interactive and digitally focused in order to get maximum impact.

Stephen: I was probably one of those football nerds at a very young age because of my Mum! I have been on other Premier League football tours through events in the tourism industry and their offerings are quite amazing, but also, they provide a way more interactive experience than just a tour around a boardroom, football pitch and changing rooms. It's about the interactivity and engagement, now more than ever.

One of the challenges you must have is that your brand is truly global. Do you find in the younger audience, particularly on educational trips or school trips, that the language is important, or more short form content? Is that something that you are continuing to evolve and change as you are developing?

Tom: There are the core education aspects which I think are really important, that we have got to make it appeal to the people who are organizing the programmes. The initial sell is to the teachers or the study support officers. I think it's all about giving them the right materials that they need before, during and after their visit. It is about how we link that into the curriculum and link into the educational needs, because it is such an important moment for young people. We can all remember how we used to love our days out with school, but if you're planning these things now it needs to have a purpose as well. It works well, because a lot of these groups come in slightly off-peak times as well. So, we have two different models. One of the models is where we can have up to 3,500 people through in a day that might not touch the sides of visitor sections. When it is that busy, we cannot have big groups going through, as it would ruin the flow of the visitor experience for the families, groups of friends or individuals. It's not self-guided, because we have lots of activities that happen along the routes, rather than having one guy taking you around. You can interact with up to 15 guides along the route which we feel is more engaging. We have also embraced the technology that is currently available, and everyone gets a multimedia guide, which is in 11 languages, soon to be 13.

We have ambitious goals, and one model we're exploring for the future is integrating educational opportunities, potentially using technology as we progress to stages two or three of our development. We're also focusing on creating programmes geared towards higher education. Recently, we hosted visits from different universities, including one from Liverpool University's tourism department. My colleague and I engaged with the students interactively, aiming to make the experience impactful. When I asked a student what their highlight was, it wasn't my presentation; it was experiencing first-hand the variety of immersive offerings we deliver.

Creating unforgettable visitor experiences

Stephen: As one of the most iconic stadiums and clubs in the world, one would think the brand would sell itself, but I guess you do have to market and sell the tourism products separately as well. How are you guys going about marketing to this younger demographic? Is there a consideration of the long-term value these young people can bring over a lifetime, alongside the importance of the short-term revenue impact? I remember the magical moment of going to Anfield for the first time. That was a long time ago, and I have continued to visit Liverpool as a club and a city, bringing friends and family from all over the world. If you get that first experience right, you will potentially have someone coming back to your brand for the rest of their life.

Tom: Yes, it is very important to have the 'wow' moment. We have created a new experience on the stadium tour where we take visitors up to the top of our main stand, to level 6, and although it is just a concourse area we have a guide sharing stories about the club up there and hit the visitors with a four-minute 'This is Liverpool Football Club' video with sound that is a really lovely, well-produced piece of content. Then we take them into the stadium and that is the first time that they see the pitch. Now, maybe it's not the same as match day but you still get that 'wow' moment. We could have cut out getting people to the very top of the stadium to see the pitch, but there's something magical about it. That's a memory which people might keep for a lifetime. So, to answer your initial question, it's about getting young people in and then creating the magical experiences that will stay with them for a lifetime.

Again, since it is the social media age, it is also about what people share when they are on the LFC experiences. There are so many brilliant photo opportunities on the tour route. However, there is a challenge on how we use the social media channels that we have at our disposal since we are still a football club first. Sharing non-football content is not what the core of the audience wants. Although we do push content, I guess for a more traditional tourism destination or venue it is a little easier to engage with young visitors' content as well as pushing out their own content.

How do we market? It is about consistently creating captivating content that visitors can share and remember.

TIP

Marketing strategies targeted at the youth market require a high degree of consistency and adaptability. Resting on past successes can lead to stagnation, as what resonates today may quickly become outdated tomorrow. To maintain a strong connection with this audience, brands must continuously refresh their approach, leveraging data, feedback and emerging platforms to stay ahead of the curve.

Stephen: I guess, since you are slightly restricted in terms of what you can push, you make sure there are experiences that those young people can share, because they become your social ambassadors with their followers. You can get to three or four thousand people without any cost in marketing, if the content is engaging and they want to share.

Tom: We do spend a lot of time and money on marketing and having the right assets. Things have changed massively with regards to how to market, especially in football. Gone are the days where we would be pushing out one of our star players at the front of one of our campaigns and as the main reason for someone to come and visit. If that player then leaves suddenly, we must change all of our marketing, whether it is physical or digital content. Whether it is billboards at airports or printed materials, once that player is gone, we must change it all. Therefore, it is important to have the right dynamic content to be able to hit lots of different markets.

Stephen: In terms of what you mentioned about the educational tools coming out, what's next ? What is your youth strategy going forwards?

Tom: It is about expanding the age groups that we are looking to reach and the different interest they may have when visiting us. In early 2025 we'll be focused on a couple of core subjects, but then we have a planning phase to follow to see what is next. Again, the aspiration from there is to see how we can impart more knowledge to people who might be here for the first time and never have experienced either a football club or a tourism venue before. So, although there are restrictions on what and when we can deliver experiences at university level due to the commitments of football matches, we would like to grow that engagement by around 30 per cent.

Social media and the youth market

Stephen: I assume it is also about reacting very quickly when things change a lot in the club. For instance, last year, Liverpool signed a player from Japan who is the national team captain and a bit of a hero in Japan. Now, I guess you would have suddenly started getting a lot of young Japanese students visiting. Obviously, you have had the same with Mo Salah, whose impact on Liverpool as a city, as much as the club, has been quite unbelievable. Particularly with the younger audience, you must be way more dynamic than 15 years ago, before social media, in terms of your offering and the assets you're talking about.

Tom: Yes, it's true, and it is interesting as there are different dynamics of what people want to absorb when they come and visit us. One of the big objectives for this year is to schedule more of our higher-value premium experiences, which are available to customers with a higher price point. One example is that you can train at Melwood (Liverpool's training ground) – you get on the team bus, have a meet and greet at Anfield, and even play a game there. This is what we would describe as a five-star day with the club. Interacting with legends of the club, live music and more, all makes this a once-in-a-lifetime experience that is so special and sticks with people. We have had to do some work in terms of what's included in these experiences, because we want to schedule more of these around match days.

Being able to deliver these experiences properly, given the restrictions of matches, and matching people's expectations, is a big challenge and something we have to overcome.

You mentioned Japan before. We don't just have a player from Japan – now two of our lead sponsors are Japanese: Japanese Airlines and Kodansha. So that shows the importance of our stature in those markets. It translates into people coming to visit the football club, and so the delivery of our stadium experience must have flexibility to ensure that wherever the guest is from they can absorb the experience directly and via mobile devices.

These commercial partners are another way of getting our brand into new markets and to entice young people to visit. We make sure we have joint layered content which has been beautifully produced to bring the surroundings to life. There is a certain dynamic from particular markets that people might want to be around, get some cool photos and say 'I was there.'

TIP

Partnering with complementary brands with different audiences to yourself is a way of reaching new markets.

Collaboration with Liverpool city region

Stephen: Liverpool as a city has got great history, from the shipbuilding to the pop culture of The Beatles and music generally, and then, obviously, the sport. You've got amazing universities such as University of Liverpool, which really is a world-leading institution. With all of that as a backdrop to the tourism product, do you think the city embraces youth, and is there a desire to attract more young people to the city?

Tom: I guess, from one perspective, having a young vibe keeps Liverpool such a vibrant city especially after 8 or 9 pm. That vibrancy doesn't happen by accident. It is because it's a place where people who are plugged in are coming to study and visit. It's a very multicultural city, and it has come on an awfully long way since I started working in the sector. I think the way that Liverpool has regenerated itself with care placemaking and using the backdrop of the history of the city to create new modern experiences has been well done. Again, it doesn't happen by accident.

When I started working for the club, and it was more looking after the event side, I would go down to London people would be saying, 'Why would we want to go to Liverpool?' Now it is not like that at all. As we said earlier, young people can amplify this through social media channels more than any marketing campaign would, so we need to keep pushing the city region to keep developing our tourism offering.

Liverpool's matured as a destination as well. There are still the groups of partygoers, stags, hens, birthday celebrations, and that is a good sign for our nightlife. But we are not as reliant on that market as I think we used to be. There has been some modelling which has been done that shows how the tourism market has matured. Looking at Millennials now and their traits and habits, what they want from a destination is very different from people of my age in their late 40s, and the city needs to adapt to that.

TIP

Making sure you are engaged with your local tourism industry is important not just for your own brand but for the whole city. Raising awareness of a destination and its importance to youth can benefit the whole tourism ecosystem.

Stephen: Do you feel that Liverpool as a city and a tourism product are all working together to try to do that, or is the focus just on luxury travel?

Tom: I think we still have a way to mature before we start focusing on luxury. Luxury is definitely not our core business at the moment, but that will come eventually, as we can see it growing already. The restaurants scene has never been as vibrant and the hotels and the bars are improving all the time. There are more dynamic and luxury and products that are getting launched than even a decade ago.

Stephen: One of the things I have found is that in the UK generally we do not necessarily focus on marketing to youth. There is an assumption that people will come, which is very dangerous, probably like Liverpool Football Club 20 years ago, where there was an assumption that, 'Oh, yeah, well, it's Liverpool, people will just come and visit.' There is now more competition and other countries' brands have changed, and happenings like Brexit haven't helped the UK's image. What I find is that there are some regions, Liverpool included, where you have got that core of the university sector. So young people are already coming through the educational side of things, including a vibrant language school sector in Liverpool now, too. After doing that for 10 or 20 years, you will start getting the reunion and nostalgia visits, where people are going to stay in luxury hotels or will want a nice restaurant meal out. I guess your development into youth products must match the desire for the city to continue to attract young people, otherwise there will be fewer young people to push as a target market for the LFC museum, for example.

Tom: I agree, and I think that is important that Liverpool as a city continues to do this. What we need to do is focus on getting the balance right. Everything goes up in price on a match day weekend, but it is about filling the rest of the time with people that want to come and visit and experience the city. We have got to make sure we continue to facilitate different markets, with different products and price points to really allow us to have a well-rounded and sustainable destination. This is not just looking at price points of hotels that go up and down depending on the demand, but making sure the rail companies have value fares especially aimed at young people. If we can make the city more accessible to young people with transport by train, we will get a lot more people visiting friends who are studying in the city. Our rail system can make last minute, on-a-whim trips so expensive that it is off-putting to more price-sensitive travellers.

If I had been sat here five years ago, I would have maybe fluffed saying we have the core offer for a younger demographic, but we do now. We have a much more dynamic offering across the city, including things like Gravity Max, a 100,000 sq ft entertainment venue, and the many museums offering great special exhibitions. We have also hosted Eurovision and that gave us amazing visibility across the globe – the feedback from those that attended was really fantastic, especially given the diversity of the audience.

A blueprint for youth-centric tourism

Stephen: One of the things that we talk about in the book is the importance of the whole tourism ecosystem. Liverpool can't have the Tower of London, but if there are no youth travel fares for train journeys to Liverpool, how will you attract young people from London?

Obviously, some countries have a better ecosystem aimed at youth than others. Even some destinations like cities do it better than others. It is great to hear that Liverpool is trying to develop out with real thought of young people in mind. It sounds like Liverpool as a city destination understands that, by developing a city-wide product that attracts young people, this will in turn create a demand for luxury restaurants and hotels in the future.

Tom: Thank you, I do hope so. A lot of my good friends came to Liverpool to study and they are still here and thriving. They are really devoted to investing in the local economy here. It is important that, as a city, we keep these bright people who come and study, so we do not have the brain drain that caused issues before. Now the infrastructure is better for visitors, it is also better for people living here and therefore people are more likely to stay and call Liverpool their home. In the early 1990s, for example, as a kid I was brought up in Liverpool and you didn't know any different. All I saw was people who were involved in big business moving out of Liverpool and not coming back. But now things are different, and we can't take that for granted.

Conclusion

When you look at a brand like Liverpool Football Club, it's not just about slapping a logo on a shirt and calling it a day. It's about the intricate balance of attracting youth while pulling them into the fold, making them lifelong fans – and, by extension, lifelong visitors to the club and the city. The trick is in the details around their marketing and experiences: short, sharp bursts of content that engage young people, paired with those 'wow' moments on their tours that leave visitors talking about it for years, or sharing them immediately on social media. If you are serious about making your attraction more appealing to the younger crowd, these are the kinds of things you need to consider.

And let's not kid ourselves (especially in English speaking countries), assuming everyone speaks the local language is a big mistake. Quality translations are not just nice to have; they are crucial, especially for your core international markets. You want the experience to resonate, to be as engaging as it possibly can be, no matter where your visitors are from.

What struck me most was how in sync the whole Liverpool city region seems to be. There is a real infrastructure in place, one that is deliberately designed to draw in young people. It is not just about slick marketing; it is about having the right products that are accessible to them. This is where public–private partnerships come into play, showing just how vital they are when you are trying to maximize a tourism destination's potential.

KEY TAKEAWAYS

1 **Youth as a strategic focus:** Liverpool FC prioritizes youth engagement, aiming to build lifelong loyalty by offering memorable, interactive experiences tailored to younger audiences.

2 **Revenue and reintegration:** The club's tourism strategy generates significant revenue reinvested into its operations, underscoring the value of a strong tourism focus, even for established sports brands.

3 **Innovative fan experiences:** Liverpool FC uses gamification and interactive tours to keep fans, especially younger ones, engaged, reflecting a shift from static displays to dynamic, tech-driven experiences.

4 **Global reach and cultural adaptation:** With a worldwide fan base, Liverpool FC adapts its tourism offerings to different cultures and languages, using global partnerships to expand its reach.

5 **Social media amplification:** By encouraging young visitors to share their experiences online, Liverpool FC leverages user-generated content as a key marketing tool to reach broader, younger audiences.

6 **Collaborative city development:** Liverpool FC works with local tourism bodies to position Liverpool as an accessible, youth-friendly destination, supporting sustainable tourism growth for the city.

16

The future of youth travel

Youth travel has emerged as a linchpin in the tourism and hospitality industry, influencing trends, driving innovation and shaping the future of global destinations. Over the course of this book we have delved into interviews with industry leaders, unpacking the evolving dynamics of youth travel. As the journey concludes, it is crucial to reflect on the insights gleaned and to consider the road ahead. Youth travellers are not just tourists; they are future ambassadors, tastemakers and long-term consumers. Their experiences today will influence their loyalty and choices tomorrow. I hope you noticed the great passion all connected with the sector have for making sure young people have the most incredible of experiences, whatever they do and wherever they go. As someone who has been fortunate enough to experience many amazing opportunities through travel, I hope this book acts as a catalyst to get brands and destinations more focused on creating these opportunities for others.

This final chapter synthesizes the key themes of the book, explores the challenges and opportunities for brands and destinations, and projects how the sector can continue to evolve to meet the needs of an increasingly informed, connected and discerning demographic.

Challenges and opportunities for the future

As we look towards the future of youth travel, several challenges and opportunities emerge that will define how brands and destinations adapt to this dynamic sector.

Gen Z and the rising Generation Alpha are digital natives, growing up in a world where technology seamlessly integrates into every aspect of life. These young travellers expect not only a presence on digital platforms but also a sophisticated mastery of engaging storytelling and innovative outreach. This necessitates a shift from traditional marketing to strategies that prioritize interaction, personalization and relevance across multiple touchpoints.

The world has never been more accessible, and destinations now compete on a global stage. Offering unique, tailored experiences has become critical to standing out. This requires deep insights into cultural nuances and consumer psychology, as well as the ability to adapt offerings to meet the expectations of a diverse audience while maintaining a distinct identity.

Technology is indispensable for engaging youth travellers, yet maintaining the authenticity and human touch of travel experiences remains essential. While virtual tours, AI-powered booking systems and immersive digital content are powerful tools, travellers still crave meaningful, real-world connections. Striking a balance between innovation and authenticity is a continuing challenge for brands and destinations.

The role of government and tourism boards is increasingly important in shaping the ecosystem for youth travel. From infrastructure investments to visa policies, collaboration with brands is essential to create seamless and appealing experiences. The youth travel sector faces a unique challenge: its oversight and regulation often fall across multiple governmental departments, depending on the purpose of the travel. Immigration, tourism and education departments all play distinct roles in shaping the policies and infrastructure that support youth mobility. This fragmentation can create obstacles for developing streamlined programmes that cater to young travellers' needs.

For youth travellers, the practical aspects of a destination are often as significant as its cultural appeal. Factors like ease of entry, safety and reliable connectivity are critical in their decision-making process. Destinations that succeed in addressing these considerations – while also providing enriching cultural experiences – stand to gain the most from this dynamic demographic.

As youth travellers come from increasingly diverse backgrounds, ensuring accessible and inclusive travel experiences is paramount. This includes addressing economic disparities, creating accommodation and attractions for travellers of all abilities, and ensuring cultural representation that resonates with a broad spectrum of identities. Forward-thinking brands and destinations must lead with inclusivity as a core value to capture the loyalty of this diverse audience.

By embracing these opportunities and tackling these challenges, brands and destinations can remain relevant, competitive and deeply connected to the ever-evolving world of youth travel.

The road ahead: Predictions and strategies

The integration of AI and predictive analytics is set to revolutionize the youth travel industry. By leveraging AI, brands can offer personalized itineraries, predict travel trends with remarkable accuracy and significantly enhance the overall customer experience. These technologies will allow companies to adapt to the unique preferences of travellers in real time, providing bespoke solutions that resonate with younger generations.

The long-term value of youth travellers has always been recognized by brands, but a recurring theme is the untapped potential for destinations to track the re-engagement of these travellers as they age. Imagine the power of a system that could monitor how often youth travellers return to a destination in later years, whether for business, leisure or with their families.

With the digitization of visas and passports, this possibility is inching closer to reality. The integration of advanced AI technologies could further enable destinations to analyse these trends with precision. Such data would not only quantify the economic and social value of youth travel but also encourage governments to invest more heavily in youth-focused travel programmes. Destinations like Canada and Australia, already trailblazers in youth mobility, could set the standard for leveraging these advancements, offering a glimpse into how the future of youth travel might unfold.

The rise of hybrid models in travel is another transformative trend. The lines between co-living spaces, educational tourism and sustainable travel are increasingly blurred, creating a landscape where destinations can cater to a wide range of multifaceted interests. This approach will offer young travellers the opportunity to combine experiences, such as learning, living and exploring in a single cohesive package. This will also allow investors and business owners in the sector to spread their risk by making sure their product is suitable for a variety of young travellers.

Health and wellbeing have become central priorities in the post-Covid travel landscape. Travellers, particularly youth, are more conscious of their physical and mental health than ever before. As a result, brands need to adapt their product by offering wellness-focused amenities and programmes that address these needs, such as yoga retreats, mental health workshops and nutritionally balanced dining options.

Young travellers are also emerging as cultural ambassadors, with travel increasingly seen as a tool for cultural exchange and social impact. Programmes promoting volunteerism, community engagement and sustainable tourism will continue to increase, encouraging youth to become active participants in transformative travel experiences that leave a positive impact on the communities they visit. Brands and destinations will need to invest in these sorts of programmes to ensure they stay relevant to youth in the future.

Finally, the importance of collaborative ecosystems cannot be overstated. The future of youth travel will rely on robust partnerships between brands, governments and local communities. These stakeholders must work together to create ecosystems that offer seamless, enriching and impactful experiences. Such collaborations will ensure that destinations and brands not only meet the expectations of young travellers but also deliver experiences that are meaningful and enduring.

Youth travel represents the nexus of opportunity and responsibility for the global tourism sector. It is a market that demands creativity, authenticity and foresight. By understanding the intrinsic and extrinsic motivations of young travellers, brands and destinations can unlock unparalleled growth and influence.

As we look to the future, the most successful players will be those who invest in creating meaningful, adaptable and sustainable experiences. The journey of youth travellers does not end with a single trip – it is the start of lifelong relationships with destinations, cultures and brands. By embracing this ethos, the industry can ensure that it remains vibrant, relevant and impactful for generations to come.

INDEX

accessibility 42–43, 47–48, 79, 187–88
acquisitions, strategy 33–37
adventure-focused hostels 47
advocacy 26, 35–36, 62–64, 75–77, 104,
    135–37, 154, 201
affordability 187–88
age brackets 199–200
ages of travellers 6–7
Air Doctor 82–83
airlines
    loyalty 18, 25
    loyalty programs 154
    marketing 30–32
    off-peak filling 25
alignment, marketing 33–37
Amazon Prime Student 31–32
ambassadors 62–64, 75–77, 104, 135–37,
    154, 201
Angkor Wat, Cambodia 48–49
annual multi-trip (AMT) insurance 80–81
anxiety 156
artificial intelligence (AI) 21, 22, 159,
    208–11, 227
Asia
    hostel brands 169
    marketing to 172–77
    traveller preferences 170–72
associated travellers 26
Athens, Greece 122–24
Australia
    ambassadors 62–64
    Covid-19 29–30
    cultural affinity 55
    data utilization 13, 18
    destination marketing 60, 63–64
    economic contributions 55–57
    as a long-haul destination 57–64
    long-term value 56–57
    positioning 53–66
    social media 61–62
    student travel 15, 18
    visa extensions 7
    working holiday scheme 7, 28–30,
        34–35, 54, 58–59, 151, 155
authenticity 30–33, 35, 63–64, 67–68,
    72–76, 79, 159–60

and technology balance 177–78
availability, of data 12–14, 20

backpackers 26
    Asia 171–72
    hostels 39–52, 166–82
Bali bombings 71–72
banking, student offers 18, 153
Battacharya, K. 39–52
Bayswater Education 127–45
blogging 39–40, 68
Bonard 11–23
Bonvoy 102–06
Bonvoy Experiences 104–06
Bournemouth, UK 141–43
brand loyalty 100–113
    digital transformation 109–10
    global vs. local 110–11
    lifetime value 111–12
    Liverpool Football Club 213–24
    programs 102–06
    social media 106–09
    targeted 101
brand positioning 33–34
brand targeting 101
Brexit 93, 95, 161
BridgeUSA initiative 151, 155
British Educational Travel Association
    (BETA) 6, 88–99
    data utilization 91
    entrepreneurship 96–97
    founding of 89–91
    global comparisons 94–96
    role of 88–89
    and UK Government 92–94
BudgetTraveller blog 39–40
BUNAC 146–49

Cambodia 48–49
Campaign Against Living Miserably
    (CALM) 203
Canada 15, 141–42
challenges 88–99, 155–63, 225–27
charitable giving 68, 72–74
children, as influencers 104
Chinese travellers 171, 173, 175

Chur, Switzerland 45–46
climate change 19
Club 18–30 6–7
clusters 190
co-creation 35–36, 61–62
co-living 183–97
  clusters 190
  see also purpose-built student
    accommodation
collaboration 73–76, 85, 107, 110–11,
  135–36, 142, 220–22
collection of data 13–14, 17, 20, 62–63
community building 39–52
  see also hostels
competition, global 81–82
contactless payments 172
content co-creation 35–36, 61–62
Contiki 7
Cope, S. 53–66
core values 32
cost of living 157–58
Covid-19
  airline demographics 25, 27
  Australia 29–30
  government support 93
  impacts 14–15, 155–58, 190–91
  and language travel 137–39
  'revenge travel' 9
  student travel 14–15
  and travel insurance 70–71
  United Kingdom 93, 95
cultural accommodations 175
cultural affinity 55, 153–54
cultural exchange programmes 146–65
  benefits of 147–49
  challenges 155–63
  cultural impacts 153–54
  economic impacts 153–54
  opportunities 155–63
  policies 150–53
Cunningham, V. 146–65
customer engagement 213–24
  beyond stays 104–06
  chat apps 176–77
  loyalty programs 102–06

data analysis 21
data availability 12–14, 20
data collection 13–14, 17, 20, 62–63,
  111–12
data utilization 12–14, 20, 91, 108–09,
  111–12, 124–25
demographics
  age brackets 199–200
  airline travel, Covid-19 25, 27
  football fans 215–16
  hostel users 168, 192–93

language travel 130–31
  of travellers 77–78
destination marketing 60, 63–64, 114–26
  Athens 122–24
  challenges and opportunities 121–22
  data-driven strategy 124–25
  ecosystems 117–18
  infrastructure 117–19
  KPIs 120–21
  long-term affinity 119–21
  new campaigns 205–06
  over-tourism 115–17
  placemaking 115
Deutschland Ticket 48
digital-first marketing/travel 30–33, 35–37,
  44, 49–50, 61–62, 96, 106–09,
  169–70, 205–09
  channels 205–06
  demographics 199–200
  platform preferences 206–09
  success factors 200–201
digital nomads 46–47, 120, 122–24
digital transformation 12–14, 20–21,
  76–78, 109–10, 138–39, 158–59,
  169–70, 185
disconnecting, technologically 178–81
diversification 25–26
dreaming 158–59
duration of travel 25–26

early impacts 132–33
economic conditions 157–58
economic contributions 16–17, 25–26,
  43–44, 55–57, 79–80, 153–54
  Australia 55–57
  multiplier effect 140, 143–44
economic slowdowns 25, 27
ecosystems
  destination marketing 117–18
  economic contributions 16–17
  housing supply 14–15, 118, 120, 192–94
  infrastructure 43, 45, 117–19, 124–25
  interconnectedness 133–37
  Liverpool 220–22
  multiplier effect 140, 143–44
  political/policy effects 118–19, 192–93
educational benefits 8–9, 216–17
e-gaming 84
emergencies 76–77
engagement
  beyond stays 104–06
  chat apps 176–77
  loyalty programs 102–06
  next generation 213–24
English, E. 88–99
enrichment, as a travel aim 19–20
entrepreneurship 96–97, 147–48

Europe, Interrail pass 47–48
evolution
    of hostels 40–41, 168
    of insurance offerings 75–78, 80
    of youth travel 6–7, 12, 75–82
experiential travel 19–20, 22, 26, 43, 47,
        104–06, 137–43, 219–20
extensions, Australian visas 7

football clubs 213–24
    demographics 215–16
    ecosystem 220–22
    educational programmes 216–17
    experience curation 217–19
    marketing 218–19
    premium experiences 219–20
Footprints 72–74
fragmentation 207

Gelsenkirchen, Germany 45
Generation Alpha 225–27
generational trends 185–88
Gen Z 8, 225–27
    authenticity 30–33, 35, 75
    e-gamers 84
    loyalty generation 19–20
    pace of change 160–61, 162
    risk aversion 156–58
Germany 44, 45, 48
global competition 81–82
global leadership programmes 130
government policies 7–8, 28–30
Greece 118, 122–24

Herbertson, J. 127–45
higher education
    banking 18
    housing supply 14–15
    value of 16
Hong Kong 61
hospitality industry 100–113
    digital transformation 109–10
    lifetime value 111–12
    loyalty programs 102–06
    social media 106–09
    targeted brands 101
Hossegor 46
hostels 39–52, 166–82
    adventure-focused 47
    Asian brands 169
    demographics 168, 192–93
    digital nomads 46–47
    digital transformation 169–70
    future trends 46–49
    reinvention 166–82
    smaller destinations 45–47
    transformation of 40–41, 168, 190–93

HostelWorld Group 166–82
hotels 100–113
    digital transformation 109–10
    lifetime value 111–12
    loyalty programs 102–06
    social media 106–09
    targeted brands 101
housing supply 14–15, 118, 120, 183–97
    see also purpose-built student
        accommodation
humour 32
hyper-connectedness 177–81

Iceland 47
impacts
    of Covid-19 14–15, 155–58, 190–91
    economic contributions 16–17, 25–26,
        43–44, 55–57, 79–80, 143–44,
        153–54
    housing supply 14–15
    long-term value 17–21
imperfection 159–60
inbound passenger surveys 17
inclusivity 226
Indonesian travellers 171
influencers 201
infrastructure 43, 45, 47, 117–19, 124–25,
        192–94, 220–22
innovation 82–85, 189–93
in-person experience 137–43, 148–49
Instagram 208
insurance 67–87
    AMT product 80–81
    collaboration 73–76
    and Covid-19 70–71
    digitization 76–78
    emergencies 76–77
    innovation 82–85
    loyalty 80–81
    perceptions of 70–71
international students 6
    banking 18
    economic contribution 16–17
    housing supply 14–15
    purpose-built accommodation 183–97
    visa changes 14–15
internships 146–65
Interrail pass 47–48, 159–60

Japanese travellers 173
JENZA 147
job market 157–58

key performance indicators (KPIs) 120–21
Korean travellers 171–72, 173

Lagom 139

language travel 127–45
    and Covid-19 137–39
    demographics 130–31
    lifelong affinity 133–37
    multiplier effect 140, 143–44
    purpose of 127–32
    work rights 131
lifelong affinity 119–21, 133–37
lifetime value 111–12
    see also long-term value
Liverpool Football Club 213–14
    educational programmes 216–17
    experience curation 217–19
    premium experiences 219–20
    regional collaboration 220–22
    role of tourism 213–14
long-haul destinations 57–64
long-term planning 188–89
long-term value 17–21, 25, 56–57, 81,
        94–96, 97, 111–12, 119–21, 131–37,
        151–53
loyalty
    airlines 18, 25, 154
    banking 18, 153
    brands 100–113
    Gen Z 19–20, 213–24
    insurance 80–81
    tracking 17–21
loyalty programs 102–06, 154
Lucey, S. 198–212
luxury products 221–22

Malaysian travellers 171
marketing 198–212
    and advocacy 26, 35–36
    alignment 33–37
    artificial intelligence 208–11
    authenticity 30–33, 35, 63–64,
        159–60
    channels 205–06
    co-creation 35–36
    constraints 203–04
    data-driven 124–25
    demographics 199–200
    destinations 60–61, 63–64, 114–26
    digital first 30–33, 35–37, 44, 49–50,
        61–62, 106–09
    football clubs 218–19
    generational divide 8
    global translation 110–11
    long-haul destinations 57–62,
        63–64
    for the long-term 201–03
    loyalty programs 102–06
    new campaigns 205–06
    platform preferences 206–09

segmentation 199–200, 209–11
    strategies 24–38, 205–06
    success factors 200–201
    targeted brands 101
    to Asian travellers 172–77
    to Gen Z 19–20, 30–33
market size 24–25
Marriott 100–113
media, and cultural affinity 55
medical emergencies 76–77
Meng, P. 166–82
mental health 178–81
Mexico 49
Midgard Base Camp, Iceland 47
Millenials 8, 72
mobile-first offerings 106, 172–73
Morro 183–84
multiplier effect 140, 143–44

national strategies 142–43
needs of travellers, hostels 41–42, 46–47,
        170–72
Netherlands, student travel 15
net promoter scores (NPS) 77

online learning 138–39
online reviews 77
outdoor activities 19, 22
over-tourism 46, 115–17

pace of change 160–61, 162
parents, influence of children 104
partner positioning 31–32, 73–76, 85, 107,
        110–11, 135–36, 220
Pavlacic, P. 11–23
peer-to-peer communication 75–77,
        154, 201
philanthropy 68, 72–74
Pinterest 208
placemaking 47, 115
planning permission 192
policy effects 118–19, 150–53, 161,
        192–94, 226
politics, impacts of 118–19
popular culture, and affinity 55
Portugal 47, 49
Pound, N. 67–87
Psarros, M. 114–26
Purnell, E. 100–101
purpose-built student accommodation
        (PBSA) 183–97
    clusters 190
    expectations 186–87
    innovation 189–93
    long-term planning 188–89
    policy effects 192–93

rail travel
    Deutschland Ticket 48
    Interrail pass 47–48
reciprocal visa agreements 58–59, 131
regional collaboration 220–22
    see also ecosystems
research
    Bonard 11–23
    data availability 12–14
    data collection 13–14, 17
    economic contributions 16–17
    long-term value 17–21
    need for 20
resilience 26–27, 28–30, 148–49
'revenge travel' 9
risk aversion 156–58
Ryanair 30–31

safety 42, 68–71, 77–79, 156–58
Scape 183–84
scholarships 68, 83
school trips, concepts 7
search journeys 31
seasonal work 146–65
segmentation 199–200, 209–11
short-termism 134
size of market 24–25
smaller destinations, hostels 45–47
Smart Delay™ 82–83
social media 30–31, 44, 49–50, 61–62,
        75–77, 96, 106–09
    and cultural affinity 55
    and experiences 219–20
    platform preferences 206–09
soft power 132–36, 151–53
South Africa 141
specified work, Australia 7
storytelling 64, 159–60
strategies
    acquisition 33–34, 205–06
    alignment 33–37
    collaboration 142
    data-driven 124–25
    marketing 24–38, 205–06
    national 142–43
student travel
    Australian framework 18
    banking 18
    Covid-19 effects 14–15
    economic contribution 16–17
    housing supply 14–15
    language travel 127–45
    purpose-built accommodation 183–97
    visa changes 14–15
Student Universe 31–32
summer camp roles 146–65
surveys, inbound 17

sustainability 19, 22, 74–75, 129, 187
Sutherland, R. 30
Switzerland 45–46

targeted brands 101
technology
    artificial intelligence 21, 22, 159,
        208–11, 227
    and authenticity 177–78
    disconnecting 178–81
    opportunities 226–27
    see also digital...
Thailand 42–44, 48–49
Thai travellers 173
TikTok 37, 50, 106–09, 120, 208–09
Toposophy 114–15
tracking, long-term loyalty 17–21, 22,
        111–12, 155
trade bodies 90
    see also British Educational Travel
        Association
transformation
    of hostels 40–41, 168, 190–93
    see also evolution
translation, local to global 110–11
transportation, Interrail pass 47–48
travel insurance 67–87
    AMT product 80–81
    collaboration 73–76
    and Covid-19 70–71
    digitization 76–78
    emergencies 76–77
    innovation 82–85
    loyalty 80–81
    perceptions of 70–71
traveller preferences, hostels 41–42,
        46–47, 170
travel with purpose 9, 19–20
travel research
    Bonard 11–12
    data availability 12–14
    data collection 13–14
    economic contributions 16–17
    long-term value 17–21
    need for 20
trust 68–72

United Kingdom (UK)
    BETA 88–99
    Bournemouth 141–43
    economic impacts 16
    failures 59–60, 95–96, 132–37, 141–42
    student travel 15, 16
United States (US)
    BridgeUSA initiative 151, 155
    economic impacts 16
    student travel 16

USIT 147
utilization, of data 12–14, 20

values, and Gen Z 32
value of youth travel 11–23, 80–81, 119–21
   economic 16–17, 25–26, 43–44, 55–57,
     79–80, 143–44, 153–54
   long-term 17–21
visa agreements 150–53
   reciprocal 58–59
   see also Working Holiday visa schemes S.
visa applications 48–49
visa changes, international students
   14–15
visa extensions, Australia 7
visitor experience curation 217–19

wellbeing 187, 189–90
Willan, S. 24–38
Working Holiday visa schemes 7–8, 28–30,
   34–35, 54, 58–59, 131, 151, 155
work of mouth 26
work rights 131
World Nomads 67–87

youth, definitions 199–200
youth-centric tourism 213–24
youth engagement 213–24
youth-focused living 183–97
   clusters 190
   expectations 186–87
   generational trends 185–88
   innovation 189–93
   long-term planning 188–89
   policy effects 192–93
youth mobility schemes 7–8
   Australia 7, 28–30, 34–35, 53–66
   policy effects 150–53, 161
   rail travel 47–48
youth travel
   definition 9
   educational benefits 8–9
   evolution of 6–7, 12
   generational trends 185–88
   government policies 7–8
   long-term value 17–21
   post-Covid 9
   power of 24–28
   value of 11–23

Looking for another book?

Explore our award-winning
books from global business
experts in Tourism, Leisure and
Hospitality

Scan the code to browse

www.koganpage.com/tlh

More from Kogan Page

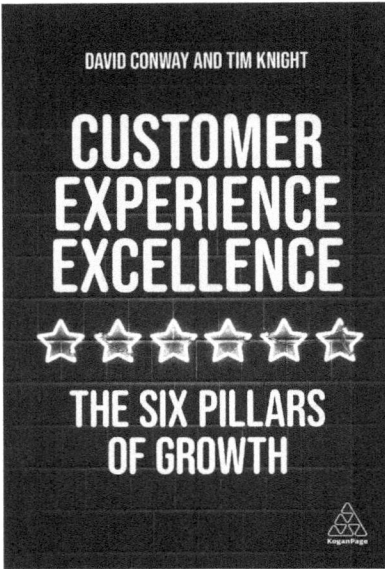

CUSTOMER EXPERIENCE EXCELLENCE

DAVID CONWAY AND TIM KNIGHT

THE SIX PILLARS OF GROWTH

ISBN: 9781398600997

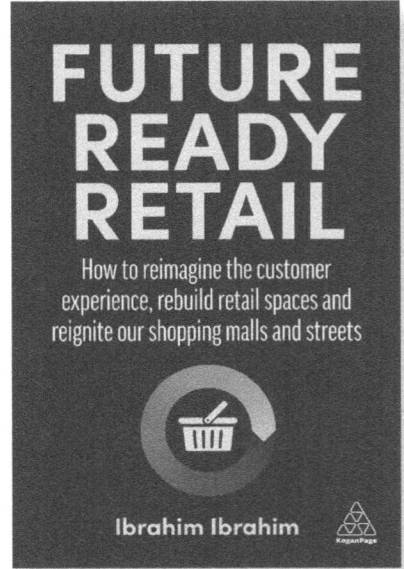

FUTURE READY RETAIL

How to reimagine the customer experience, rebuild retail spaces and reignite our shopping malls and streets

Ibrahim Ibrahim

ISBN: 9781398603349

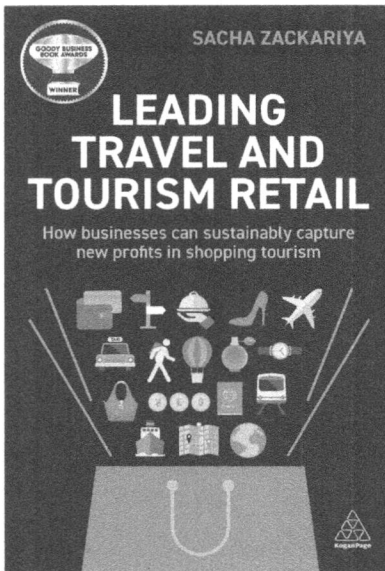

LEADING TRAVEL AND TOURISM RETAIL

SACHA ZACKARIYA

How businesses can sustainably capture new profits in shopping tourism

ISBN: 9781398609501

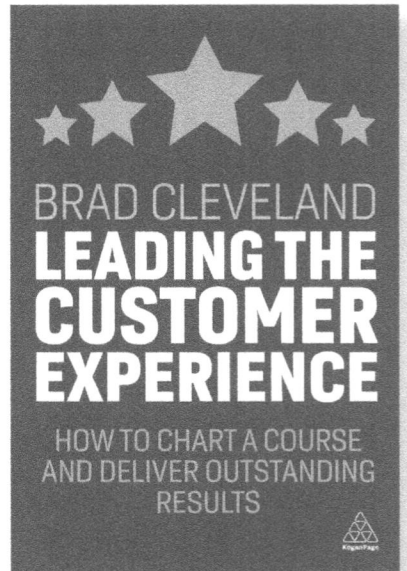

LEADING THE CUSTOMER EXPERIENCE

BRAD CLEVELAND

HOW TO CHART A COURSE AND DELIVER OUTSTANDING RESULTS

ISBN: 9781789666878

www.koganpage.com

From 4 December 2025 the EU Responsible Person (GPSR) is:
eucomply oÜ, Pärnu mnt. 139b – 14, 11317 Tallinn, Estonia
www.eucompliancepartner.com

www.ingramcontent.com/pod-product-compliance
Lightning Source LLC
Chambersburg PA
CBHW040916210326
41597CB00030B/5101